AGRICULTURAL ISSUES AND POLICIES

AGRICULTURE, FOOD, AND FOOD SECURITY

SOME CONTEMPORARY GLOBAL ISSUES

AGRICULTURAL ISSUES AND POLICIES

Additional books in this series can be found on Nova's website under the Series tab.

Additional e-books in this series can be found on Nova's website under the e-Books tab.

AGRICULTURAL ISSUES AND POLICIES

AGRICULTURE, FOOD, AND FOOD SECURITY

SOME CONTEMPORARY GLOBAL ISSUES

CLINTON LLOYD BECKFORD
EDITOR

Copyright © 2018 by Nova Science Publishers, Inc.

All rights reserved. No part of this book may be reproduced, stored in a retrieval system or transmitted in any form or by any means: electronic, electrostatic, magnetic, tape, mechanical photocopying, recording or otherwise without the written permission of the Publisher.

We have partnered with Copyright Clearance Center to make it easy for you to obtain permissions to reuse content from this publication. Simply navigate to this publication's page on Nova's website and locate the "Get Permission" button below the title description. This button is linked directly to the title's permission page on copyright.com. Alternatively, you can visit copyright.com and search by title, ISBN, or ISSN.

For further questions about using the service on copyright.com, please contact:
Copyright Clearance Center
Phone: +1-(978) 750-8400   Fax: +1-(978) 750-4470   E-mail: info@copyright.com.

NOTICE TO THE READER

The Publisher has taken reasonable care in the preparation of this book, but makes no expressed or implied warranty of any kind and assumes no responsibility for any errors or omissions. No liability is assumed for incidental or consequential damages in connection with or arising out of information contained in this book. The Publisher shall not be liable for any special, consequential, or exemplary damages resulting, in whole or in part, from the readers' use of, or reliance upon, this material. Any parts of this book based on government reports are so indicated and copyright is claimed for those parts to the extent applicable to compilations of such works.

Independent verification should be sought for any data, advice or recommendations contained in this book. In addition, no responsibility is assumed by the publisher for any injury and/or damage to persons or property arising from any methods, products, instructions, ideas or otherwise contained in this publication.

This publication is designed to provide accurate and authoritative information with regard to the subject matter covered herein. It is sold with the clear understanding that the Publisher is not engaged in rendering legal or any other professional services. If legal or any other expert assistance is required, the services of a competent person should be sought. FROM A DECLARATION OF PARTICIPANTS JOINTLY ADOPTED BY A COMMITTEE OF THE AMERICAN BAR ASSOCIATION AND A COMMITTEE OF PUBLISHERS.

Additional color graphics may be available in the e-book version of this book.

Library of Congress Cataloging-in-Publication Data

ISBN: 978-1-53613-483-4

Published by Nova Science Publishers, Inc. † New York

This is an excellent book, and will be an especially stimulating source of information for students and scholars interested in food and agriculture. The editor has assembled an impressive group of scholars, all experts in their chosen fields. The volume's strength lies in the way each chapter offers a fresh, insightful and evidence-based appraisal of some of the most pressing issues confronting the global agriculture sector today. The book also offers several key recommendations for enhancing and transforming agricultural production in the global South, which could provide a vital resource for the policy and research community.

Dr. Kevon Rhiney, Department of Geography, Rutgers University

CONTENTS

Preface		ix
List of Figures		xix
List of Tables		xxi
Chapter 1	Harnessing the Nutritional and Commercial Benefits of Jackfruit (*Artocarpus heterophyllus*) in the Tropics and Subtropics *Noureddine Benkeblia*	1
Chapter 2	Influencing Factors on the Iodine Content of Food: An Introductory Review of Non-Animal I-Sources *Gerhard Flachowsky*	29
Chapter 3	Simulating the Future of Food Deserts in Ypsilanti, Michigan Using Markov Chains and Cellular Automata *Hugh Semple*	59
Chapter 4	Influencing Factors on the Iodine Content of Food: A Review of Animal Sources of Iodine *Gerhard Flachowsky, Ulrich Meyer, Ingrid Halle and Andreas Berk*	81

Chapter 5	Unlocking the Full Potential of Carambola (*Averrhoa carambola*) as a Food Source: Botany, Growing, Physiology and Postharvest Technology *Noureddine Benkeblia*	**131**
Chapter 6	Food and Nutrition Security through Urban Food Gardens: The Roles and Potential of Community Gardening *Clinton Beckford and Blessing Igbokwe*	**159**
Chapter 7	Assessing Vulnerability to Climate Change Across Agroecological Zones in Jamaica: Insights for Community-Based Adaptation Strategies *Donovan Campbell*	**189**
List of Contributors		**223**
Index		**227**

PREFACE

Issues of agriculture, food and food security, are inextricably linked in complex and multidimensional ways. On a global basis, the role of agriculture as an economic activity has declined in terms of its contribution to Gross Domestic Production (GDP) and the proportion of the labour force employed. Even in economies that are still in the early stages of economic transition from 'traditional' to modern economies, the economic role of agriculture has diminished. Despite this trend, agriculture as an economic activity remains central to human health and existence, and food and food security issues as a whole, constitute one of the most pressing global challenges.

In 1970, Nobel Prize laureate Norman Borlaug, celebrated for his role in initiating the Green Revolution in agriculture, spoke to the dire threat he perceived the world to be facing- that of a burgeoning population and the implication that this could have for the ability of the world to feed its population. Borlaug reasoned then, that at the prevailing rate of growth, the world's population would reach 6.5 billion by the year 2000 which he thought could have serious ramifications for the 'battle against hunger' if the world became complacent. Borlaug was not far off the mark in his projection of population and growth and some would argue that his warning was manifested in the world food crisis of 2008, when dramatic increases in price of critical cereals-triple in the case of rice and double in the case of maize and wheat- resulted in social unrest which at its worse was manifested in food riots in some twenty countries.

While the situation has abated, the world as a whole and many individual nations, still face significant food challenges. In mid-2017, the world population reached 7.5 billion. It is believed that as many as 2 billion of these people suffer from hunger. This includes some 800 million people considered to be chronically undernourished and over a billion more who experience 'hidden hunger'- a term used to describe micronutrient deficiency in humans. The relationship between population growth and food supply is neither linear, nor simple. Some people have described the current world food problem as a paradox of hunger in the midst of plenty. Scientific and technological advances in agriculture have resulted in significant increases in food production which has kept pace with population increase. Today the world produces more than enough food to feed its 7.5 billion population.

Yet paradoxically, an estimated 800 million people worldwide are severely undernourished. The problem is complicated. First, food production globally is uneven. There are countries and areas of food surplus and countries and areas with food deficit creating what we might see as oases of food on the one hand, and food deserts on the other. This situation structures itself according to economic fortunes in the Global North and the Global South respectively. Secondly, the machinations of the global politically economy have created realities and structural problems which make food distribution and transfer from surplus areas to deficit areas, challenging if not prohibitive. The reality is that it is often more economically feasible to dump surplus food produced in developed countries. Each year millions of tons of food end up in landfills in the Global North while hundreds of millions of people in the Global South- and to be fair, a growing number in the Global North- experience hunger. Combine this with issues of access- what Nobel Prize laureate Amartya Sen called 'entitlements' (Sen 1981)- in many economically disadvantaged nations and the analysis becomes even more dire.

The prognosis for the future provides cold comfort. By the end of this millennium, global population is projected to hit the 10 billion mark. So, the question is, how will world provide adequate food and nutrition for 10 billion people in ways that are ethical, safe, equitable and sustainable. The glib answer might be that it has been done before and we simply need a new Green Revolution. This answer would neglect the current context-a complex and challenging agro-ecological milieu in which food politics demands that food production be more ethical, more natural, and more sustainable. The reality then, is that we must now do more with less. Produce more food that is healthier and more nutritious. We must do so in ways that preserve environmental integrity and maintain bio-diversity. We will have to do this with less land and less chemical inputs.

More significantly, we must do so in the context of global climate change and variability. In 1970, Borlaug referred to the 'population monster' in assessing the problem of feeding the world. Today, he might see climate change in the same light. Obviously, the precise impact of climate change on agriculture and food production cannot be known with full certainty at this time. However, the prognosis is for far reaching consequences and impacts- positive and adverse- which will be uneven and differential. Projected negative manifestations of climate change include warming temperature, sea level rise, increasingly unpredictable weather, and increased intensity of extreme meteorological events. Some areas will experience salinity, while others will face desertification. Water resources could be affected with serious implications for agriculture and hence, food security. Climate change it seems, will also force us to have to grow more food, with less land and less water. Impacts on agricultural diseases and pests are also expected as there is already evidence of diseases and pests adapting to exploit changing climatic conditions (CGIAR/CCAFS 2014; University of California, Davis 2015; Porter et al. 1991).

Research indicates that there is growing evidence that the effects of climate change are already being felt in many places, and there is a global scramble to develop and implement strategies for building resilience and adaptation to climate change and climate variability. 'Climate smart' agriculture, will be a big part of the any new Green Revolution.

As the global population grows, the attention to food security issues will become heightened. Food security refers to a situation where people have access to adequate, healthy and affordable food. This means that food is *available, accessible* in the sense of people being able to purchase it, *safe* in that it is healthy and pose no harm for human consumption, and *nutritious* in that people are eating 'good' food that provide the micronutrients and energy they need to lead active and productive lifestyles. The multiple dimensions of food security underscore the complexity of the global food security challenge and the need for nuanced responses. The problem has to be addressed in the macro that is, global context through changes in the global political economy in which the processes of globalization work to provide global food networks with mechanisms for a more efficient and ethical distribution of food from surplus areas to deficit areas. This has to be done with concurrent strategies at the micro level. The question here is, how do we increase food availability and access of safe healthy and nutritious food in food deficit areas to make them more self-sufficient in food?

Globalization has contributed to a decline in food production in many parts of the global south and a concomitant increase in food imports from food surplus areas. In essence, what happens is that food produced with the benefit of generous subsidies in countries in the Global North is marketed in poor countries of the Global South. These cheap foods flood local markets making local agricultural produce uncompetitive and therefore making local food production uncompetitive. This has become a major disincentive to local agriculture which is stagnating in many tropical countries and declining in others and leads to increased food importation to make up food deficits. The result is that many tropical countries have transitioned from being net producers of food, to being net importers of food. For example, Africa has a whole has become a net importer of food in the last 40 years or so (Rakotoarisoa, MA, Iafrate, M & Paschale, M 2012).

The overall effect of this growing reliance on food importation is to make food security in these countries at best, precarious, as dependence on external food sources put local food supplies at the vagaries of the international political economy and increases susceptibility to external shocks. That being said, it is important to emphasize that large dependence on food importation does not necessarily means that a country is food insecure. In fact, many of the biggest importers of food are also some of the richest countries in the world. These are led by the United States and include countries like Germany, Japan, France, Italy, and the United Kingdom. These countries are not food insecure and could be completely self-sufficient if they wanted to be. Importing some foods from other places simply makes economic sense and provides variety and diversity in food choices.

Borlaug in the 1970s, suggested that the global food supply could be secured if the full potential of science and technology was unleashed without hesitation. This sentiment is equated with advocacy for approaches that facilitate the mass industrial production of food and widespread use of genetically modified foods. This approach no doubt favours large multinational agricultural corporations, which are the ones who can afford GMOs. The vast majority of small-scale food farmers who are the backbone of food production in developing countries are not able to access expensive GM technologies. The point has to be made that mass food production has not solved the problem of global hunger, although it has dramatically increased food production globally. What it has done is increased the gap between rich and poor countries in terms of food production and heightened the terms of trade imbalances between these two groups of countries. GM and other industrial agricultural technologies have had no positive effects on alleviating food security issues in food deserts and food deficit areas.

Furthermore, the argument could be made that the full weight of science and technology has been unleashed and is receiving considerable critique in terms of its environmental consequences, its cost, and its impact on human health. There is a growing global trend away from this kind of food production and a gravitation towards more 'natural' and 'wholesome' foods produced locally. As Clinton Beckford discusses in Chapter ..., there is growing skepticism about the benefits of GM foods and concerns about their economic, environmental and human health impacts.

This brings us to another irony. While some people in some parts of the world struggle to eat enough and suffer from symptoms of hunger, others in other parts of the world suffer from another set of diet related diseases. Chronic diseases related to food consumption are on the rise in many industrialized societies including obesity, hypertension, diabetes, and heart disease. These diseases in the context of food, are a function of malnutrition-not undernourishment. So, while hunger, which is related to undernourishment, is most associated with underdeveloped countries, developed countries have their own problems related to nutrition, which GM foods have not solved, and some observers think might actually contribute to. It is important to point out too, that chronic diseases such as obesity, diabetes, hypertension, and heart disease, are also on the increase in developing countries where there has been a steady increase in the availability and consumption of industrial produced foods as people become estranged from their local and traditional foods.

In addition to the .8 billion people suffering from chronic hunger, another 1.2 billion suffer from micronutrient deficiency. This adds another layer to the extent of the global food challenge. The quality of food being consumed, is as big a challenge as the quantity. Solving the global food and nutrition problem then, will require more than just increasing the global production of food. Multiple approaches are needed and improving food production in areas of food deficit is urgent and vital.

This book looks at agriculture and food in a framework of enhancing food security. The seven chapters that comprise the volume are written by successful international scholars with sterling academic publication records. The authors live and work in countries across the world including Jamaica, Germany, Canada, and the United States. Together they have conducted numerous research projects and published a high volume of scholarly articles on issues related to the central themes of this book-agriculture, food and food security. The chapters in this volume cover major issues such as increasing food availability, enhancing micronutrients, localizing food production, maximizing the food and nutrition potential of fruits, addressing food insecurity, and vulnerability to climate change and climate variability.

In Chapters 1 and 5, Noureddine Benkeblia discusses the role and potential of two tropical fruits-star fruit and jackfruit- to enhance food and nutrition security. Benkeblia argues that these two fruits have significant potential to contribute to food availability as well as increased agricultural production through commercial farming. He posits that the potential of these two fruits is not being realized because of limited research and scientific development and hence knowledge, about their value.

Preface

In Chapter One, Benkeblia shows that the composition of jackfruit is comparable with various other commercial fruits and might be a good source of nutrients which could contribute to improving the nutritional status of rural populations in the tropics where the fruit grows well. The chapter also discusses the potential of jackfruit as a source of income for growers and the possibilities for integration in traditional agroforestry systems producing economic benefits for poor farmers. This chapter focuses on critical issues relevant to the production, utilization and postharvest handling of jackfruit in order to highlight the botanical features of the plant that have implications for assessing its nutritional and commercial value.

In Chapter Five, Benkeblia turns the spotlight on Carambola. This fruit is produced in large orchards in USA, Israel, India, Malaysia, Australia and many other countries. In the Caribbean and other tropical regions but mainly as isolated trees and only rarely in orchard-like stands. The fruit is known and eaten but, not like other tropical fruits like orange, banana, mango, melon, papaya, pineapple to name some of the most popular fruit types. The author posits that the nutritional and commercial potential of carambola fruit has been barely tapped and much more work is required on various aspects of its physiology and biochemistry. To this end the chapter explores salient issues of production and processing that should be considered in exploiting the fruit for nutritional and commercial purposes. These discussions have implications for tropical and subtropical areas in Africa and Asia where many people experience food insecurity. In the Caribbean, where both these fruits grow extremely well, their potential as food sources and employment and income generators have not even begun to scratch the surface.

Gerhard Flachowsky, examines issues of micronutrient intake with specific reference to iodine intake in humans. Iodine (I) is an essential trace element for humans and animals. Two centuries of iodine research have shown many interesting results, but also opened some new questions in animal and human nutrition, which Flachowsky seeks to address. Iodine is part of the thyroid hormones triiodothyronine (T3) and thyroxine (T4) and therefore, it is highly important for key processes in the body. For example, a lack of iodine is known to cause a loss of viability in embryos, brain development delay, enlargement of the thyroid gland a condition known as goiter, and severe iodine deficiency during pregnancy is known to cause mental retardation. Other health consequences related to iodine deficiency, include hypothyroidism and iodine induced hyperthyroidism Iodine deficiency still remains a major public health issue in many countries, including some European countries. World Health Organization (WHO) estimates that about 800 million people would suffer from iodine undersupply but altogether the number of countries with insufficient supply of iodine is decreasing- from 54 in 2003 to 30 in 2012. On the other hand, excessive iodine intake has been recorded in ten countries.

In Chapter Two, Flachowsky explores issues related to iodine from plant sources. In Chapter Four, he collaborates with colleagues Ulrich Mayer, Ingrid Halle and Andreas Berk, to examine critical issues related to human iodine intake from animal sources. In both cases, the authors dive deeply into a plethora of issues related to iodine sources, availability and access, and the implications for human nutrition and health.

In Chapter Three, Hugh Semple explores the growing phenomenon of food deserts in the Ypsilanti area of Michigan in the United States. Food deserts are communities and neighbourhoods mainly found in urban areas that are characterized by high levels of food insecurity. They are low-income communities which have limited access to affordable, nutritious and healthy foods, largely because of the lack of large supermarkets that carry a wide variety of foods including fresh fruit and vegetables, and they have a preponderance of convenience store types of facilities that carry a limited range of foods. They also have a lot of fast food facilities. Residents are required to travel outside their communities but often depend on unreliable and expensive public transportation. The author uses Markov Chains and Cellular Automata to analyze the growth of food deserts and predict future trends in their occurrence in the study area. Their findings show that based on socio-economic projections, we can expect a moderate increase in food deserts in the short to medium term.

In Chapter Six, Clinton Beckford and Blessing Igbokwe discuss the role and potential of community gardens to address critical issues of food and nutrition security. The chapter focuses on the potential of community gardens to enhance food and nutrition security and address obesity and other chronic diseases in low-income urban neighbourhoods by improving access to, and consumption of, healthful and affordable fresh fruits and vegetables. The authors argue that promoting community gardening as part of a larger urban poverty alleviation strategy anchored in a robust food security strategy, can bring economic, environmental and health benefits to urban households and communities. They posit, that while there is some literature about urban gardening in general, there is a lack of research and scientific literature specifically about community gardening in the developing world and especially in parts of food deficit areas in the tropics and sub-tropics where community gardens could be a vital element of agricultural renewal and household food security for both urban and rural peoples. The point is made, that community gardening is likely to become more popular as awareness of their potential and advantages increases, and consumers' demand fresh foods that are produced closer to home and are produced under more 'ethical' conditions that have implications for the environment and human health.

In Chapter Seven, Donovan Campbell explores vulnerability of farming communities in Jamaica to climate change and variability. Campbell, utilizes a livelihood vulnerability assessment approach, to identify entry points for building climate resilience in selected farming communities across nine agro-ecological zones in the small island state. The vulnerability assessment, is based on a survey of farming households (n=618) and focus group discussions in each community to assess levels of livelihood exposure to climate variability and change, climate impacts, and adaptive capacity. Campbell posits that while community-based vulnerability assessments have become more popular in Jamaica, and has illuminated important micro-scale issues, integrated macro-scale analyses are generally lacking. Agriculture is one of the most vulnerable sectors to the impacts of climate change. Dr. Campbell, argues that the current vulnerability assessments through traditional fragmented sectoral methods are insufficient to capture the effects on complex agricultural systems. Therefore, he advocates that traditional methods need to be replaced with more integrated approaches. The goal of the study reported in this chapter, was to propose a holistic vulnerability assessment method for agricultural systems in Jamaica. The author concludes, that the findings of this comprehensive assessment, underscore the place and contextual nature of vulnerability and demonstrate how this information can be used to shape current and future climate resilient strategies by aligning community needs with adaptation priorities.

The chapters in the book provide insights into several key themes-food, agriculture and food and nutrition security. The volume explores some pressing issues including food deserts, micro-nutrient deficiency, increasing food production by tapping into the potential of already known species, localizing food production, and the agricultural vulnerability if the context of one of the most pressing global threats- climate change and variability.

I hope you find it useful and informative to your teaching and research endeavours.

Clinton L. Beckford (PhD)
University of Windsor
Windsor, Ontario, Canada

LIST OF FIGURES

Figure 1.1.	(A) Jackfruit tree (B) ripe fruit on the tree and (C) transcersal section of a ripe fruit.	4
Figure 1.2.	The main volatile compounds found in jackfruit	7
Figure 3.1.	Two-state Markov Model.	65
Figure 3.2.	Actual and Simulated Food Deserts for 2010 based on 1990-2000 Transition Matrix.	73
Figure 3.3.	Simulated Food Deserts, 2020 and 2030 based on 1990-2000 Transition Matrix.	73
Figure 4.1.	Dependence of the milk iodine concentration on the kind of iodine supplementation in diets without and with iodine antagonists via rape seed meal (RSM; n = 8, ▲, control/iodide; △, control/iodate; ■, RSM/iodide; □, RSM/iodate.	86
Figure 5.1.	Carambola (Averrhoa carambola) tree (A), leaves and flowers (B), and ripe fruits (C).	135
Figure 5.2.	Internal and external browning of carambola harvested ripe and stored at 5°C for 10 days. Left: cross section of a fruit with arrow indicating areas of chilling injury. Right: fruit with external chilling injury observed.	148

List of Figures

Figure 7.1.	Location of communities in relation to agro-climatic zones.	**197**
Figure 7.2.	Patterns of adaptive capacity, exposure-sensitivity and vulnerability.	**202**
Figure 7.3.	Spatial pattern of community vulnerability.	**203**
Figure 7.4.	Spatial pattern of community exposure-sensitivity.	**204**
Figure 7.5.	Spatial pattern of community adaptive capacity.	**205**
Figure 7.6.	Socio-demographics.	**207**
Figure 7.7.	Social network & organization.	**207**
Figure 7.8.	Knowledge & Awareness.	**208**
Figure 7.9.	Asset-base.	**208**
Figure 7.10.	Biophysical determinants.	**209**
Figure 7.1.	Socio-economic drivers.	**209**

LIST OF TABLES

Table 1.2.1.	Nutritional composition of jackfruit pulp	6
Table 1.2.2.	Changes in morphological character of fruit, bulb and seed at 15-day intervals from fruit set	13
Table 2.2.1.	Recommended iodine intake of humans (µg/day) by various scientific bodies	33
Table 2.4.1.	Water iodine levels by various authors	37
Table 2.4.2.	Iodine content of forages (e.g., grass, hay, straw, silages etc.) by various authors	40
Table 2.4.3.	Influence of plant species and vegetation stadium of iodine concentration (µg/kg DM) of various feeds in Germany	41
Table 2.4.4.	Iodine content of cereals and cereal co-products by various authors	41
Table 2.4.5.	Iodine content of legumes and their by-products by various authors	43
Table 2.4.6.	Iodine content of fruits and vegetables	44
Table 3.4.1.	Transition Probability Matrix	67
Table 3.5.2.	Predicted Probabilities for Food Desert and Non-Food Desert Census Tracts	72
Table 3.5.3.	Various Kappa Indexes	74

Table 4.2.1.	Mean iodine content in fresh milk weight samples (µg/L) by various authors	83
Table 4.2.2.	Influence of iodine source potassium iodide and calcium iodate hexahydrate on iodine concentration of milk (µg per liter)	85
Table 4.2.3.	Influence of various amounts of crambe cake (50.4) and crambe meal (77.4 mmol glucosinolates/kg DM) in rations of dairy cows on the iodine concentration of milk (n = 10 cows; 0.8 mg iodine/kg DM)	87
Table 4.2.4.	Influence of summer (outdoor, grazing) and winter (indoor) animal feeding and keeping on the iodine concentration of bulk milk (µg/kg) in some European studies	88
Table 4.2.5.	Influence of type of farming on the iodine concentration of bulk milk (µg/kg) in some European studies	90
Table 4.3.1.	Iodine content in poultry eggs presented by various authors and food composition tables	94
Table 4.3.2.	Influence of various I-supplementations (KIO_3) for laying hens on the I-concentration in whole eggs, egg yolk and egg white as well as I-recovery in whole eggs (n = 3)	96
Table 4.3.3.	Mean iodine concentrations of different samples (fresh matter) after $Ca(IO_3)_2$ supplementation (n = 6) on iodine concentration of yolk, albumin and whole egg	96
Table 4.3.4.	Iodine concentration of eggs (µg/kg complete eggs) depending on days of experimentation and level of iodine supplementation	97
Table 4.3.5.	Iodine concentration of eggs (µg/kg complete eggs) depending on days of experimentation and level of iodine supplementation from KI and $Ca(IO_3)_2$	100

List of Tables

Table 4.3.6.	Iodine concentration of eggs (µg/kg complete eggs) depending on days of experimentation, the level of iodine supplementation and diets without or with 10% rape seed cake- 24 eggs per group	101
Table 4.3.7.	Comparative analysis of iodine concentration of LSL and LB hen breeds	102
Table 4.3.8.	Iodine content in samples (µg/kg) of white and yolk fraction of eggs from free range and indoor keeping	104
Table 4.3.9.	Carry over values of iodine from feed to eggs (% of added iodine in eggs)	105
Table 4.3.10.	Contribution of daily consumed eggs to the daily iodine intake of consumers depending on iodine supplementation of the hen feed	106
Table 4.4.1.	Iodine content of meat and meat products by various authors	108
Table 4.4.2.	Influence of I-supplementation to complete feed (KI and $Ca(IO_3)_2$) on I-content of body samples (µg/kg) of broilers	109
Table 4.4.3.	Influence of I-supplementation to complete feed on I-content of body samples of growing/fattening pigs	109
Table 4.4.4.	Influence of I-supplementation of feed (concentrate and maize silage) on I-content of body samples of growing/fattening bulls	109
Table 4.5.1.	Iodine content of fish and fish products by various authors and food value tables	110
Table 5.1.	Chemical composition of carambola fruit	139
Table 7.2.1.	Vulnerability indicators	196
Table 7.2.2.	Community sample size, agroecological zone and parish	198
Table 7.3.1.	Vulnerability component scores at the community level	200

In: Agriculture, Food, and Food Security
Editor: Clinton Lloyd Beckford
ISBN: 978-1-53613-483-4
© 2018 Nova Science Publishers, Inc.

Chapter 1

HARNESSING THE NUTRITIONAL AND COMMERCIAL BENEFITS OF JACKFRUIT (*ARTOCARPUS HETEROPHYLLUS*) IN THE TROPICS AND SUBTROPICS

*Noureddine Benkeblia**
Department of Life Sciences/Biotechnology Centre,
University of the West Indies, Mona Campus, Kingston, Jamaica

1. BACKGROUND

Tropical crops, including fruits, play an important role in the daily diets of not only local populations, but also billions of people around the world. Many tropical fruits are harvested either from wild or locally cultivated trees of a wide range of minor species. Although not well known in north America and Europe, jackfruit (*Artocarpus heterophyllus*), is widely consumed in the Caribbean, and Asia where it originated. However, this fruit remains still

---

* Corresponding author email: noureddine.benkeblia@uwimona.edu.jm.

underutilized because it has undergone only limited scientific improvements and research, as is the case with many other tropical fruits. Many studies showed that the composition of jackfruit is comparable with various other commercial fruits and might be a good source of nutrients and therefore contributing in improving the nutritional status of rural population. The studies also show that jackfruit has commercial potential as a source of income for growers. The tree might also be adopted in the traditional agroforestry systems producing economic benefits for poor farmers. This chapter aims to elucidate some issues relevant to the production, utilization and postharvest handling of jackfruit in order to highlight the botanical features of the plant that have implications for assessing its nutritional and commercial value.

2. INTRODUCTION

2.1. Origin, Botany, and Morphology

The Jackfruit (*Artocarpus heterophyllus*) is a species of tree that belongs to the genus *Artocarpus* and is a member of the Moracaeae family. This botanical family consists of nearly fifty species of trees, all natives of Southeast Asia and the Pacific Islands (Purseglove 1974). The genus *Artocarpus* comprises about 50 species of evergreen and deciduous trees (Jagtap & Bapat 2010a), with three very well-known members of the genus cultivated and utilized throughout the Caribbean and Asia: breadnut (*A. camansi*); breadfruit (*A. altilis*); and jackfruit (*A. heterophyllus*). Today, the jackfruit occupies a range that extends through central and eastern Africa, throughout Asia, the Caribbean, Florida, Brazil, Australia and the Pacific Islands (Eleviteh et al. 2006).

The jackfruit is adapted to humid tropical and sub-tropical climates. It thrives from sea level to an altitude of 1600 m. It is adaptable with a growing range that extends into much drier and cooler climates than the other *Artocarpus* species (Popenoe 1974). For optimum production, jackfruit requires warm, humid climates with evenly distributed rainfall of 1000-2400 mm per annum. The tree bears good crop particularly between latitudes of up to 25° north and south of the equator (Soepadmo 1991). The jackfruit tree prefers well drained soils but can be cultivated on a variety of soils including deep alluvial, sandy loam, or clay loam of medium fertility, calcareous or lateritic soil, and shallow limestone or stony soil with pH of 5.0-7.5 (Haq 2006).

The leaves are generally oblong or elliptic though those on upper branches tend to be more obovate and those on young shoots more oblong and narrower. Leaves may reach lengths of 25 cm long and 12 cm wide at or just above the middle of the tree where they are broadest. The lamina is coriaceous, stiff and dark shiny green above and pale green beneath. Venation is pinnate 5-8 pairs of veins. Leaves are cuneate or obtuse at the base but can be irregular shaped on young plants. The lamina is flat, wrinkled or with upwind sides. The leaves are arranged alternately on horizontal branches, but tend to be spiral on ascending branches with 2/5 phyllotaxis.

Jackfruit is anemophilous, with a solitary inflorescence and both male and female produced separately on short axillary leaf twigs, either on the tree trunk or on the older branches. The individual flowers are borne on an elongated axis and are grouped into a racemoid inflorescence. The female spikes are borne on footstalks, while the male spikes are both on the footstalks as well as on the terminal shoots. One male inflorescence is formed per terminal shoot in each flowering season (Manalo 1986). The male inflorescence together with the bud and the leaf primordial is sheathed with stipules. As the bud grows larger the stipules open to expose the bud, a new leaf and the spike. At emergence, the male spike is 3-3.5 cm long and 1.5 cm wide but increases in size, assuming an oblong clavate shape and reaching 5-10 cm in length and 2-3 cm in width. There are no interfloral bracts (Moncur 1985).

The female inflorescence is 4-15 cm long and usually found distal to the male inflorescence. They tend to be more cylindrical or oblong than the male. Interfloral bracts are present. The female spikes are composite, large bright green and have a segmented surface (Alexander et al. 1983). Staminate spikes are produced on the terminal leaf axil and on the footstalks emerging from primary and secondary branches. When young, the spikes are enclosed in thick, leathery and deciduous spathes.

Figure 1.1. (A) Jackfruit tree (B) ripe fruit on the tree and (C) transcersal section of a ripe fruit.

Jackfruit is a syncarp (multiple fruit) with green to yellow-brown syncarp 30-100 cm long and 25-50 cm in diameter (Figure 1.1). The heavy round cylindrical to pear shaped fruit held together by a central fibrous core, which usually weighs 4.5-30 kg, although a weight of 50 kg has been recorded (Morton 1987). The exterior rind is warty with numerous pyramidal sections, bluntly conical carpel apices that cover a thick, rubbery, whitish to yellow wall. The perianths of the individual flowers become the fleshy pericarp surrounding the seeds. The pericarp is yellow-white or yellow and waxy firm (Corner 1938).

There are two cultivars of jackfruit that are recognized, one with firm, sweet pulp and the other with tender (or softer), mushy pulp. The former cultivar is considered to be superior. Jacob and Narasimhan (1992), reported jackfruit consist 30% of edible portion (pulp), 12% seed and about 50% rind (waste) on a fresh weight basis. The fresh de-seeded sweet pulp is consumed by people but cannot be stored for a long-time due to its high perishable nature. Attempts have been made to process jackfruit pulp, seeds and rind into various products like canned jackfruit, pulp in syrup, raw jackfruit pickle, jackfruit leather, canned seeds in brine, roasted jackfruit seeds and jackfruit seeds flour (Munishamanna et al. 2007). However, processing of jackfruit is still limited because relatively little is known about this fruit compared to many other tropical fruits (Mondal et al. 2013).

2.2. Composition and Nutritional Value

Depending on the type or clone, jackfruit is consumed fresh, or as canned slices, fruit juice and dried chips. For fresh consumption, the fruit is commonly sold as whole fruit, sections/quarters, and in some countries as minimally processed (MP) produce. The demand for marketing jackfruit as MP product has been expanding during the last 20 years because of new trends of consumers and the increasing demand for the ready-to- eat fresh produce (Jagadeesh et al. 2007; Punan et al. 2000).

The energy available to humans from ripe jackfruit has been calculated at approximately 2 MJ kg^{-1} fresh weight ripe perianth (Ahmed et al. 1986). The jackfruit pulp is a good source of carbohydrates, vitamin A and a fair protein source (Narasimham 1990). The nutritional composition of jackfruit pulp typically found in Malaysia, one of the major growers in the world, is given in Table 1. The Malaysia jackfruit pulp analyzed by Tee et al. (1997) was high in carbohydrates, fiber, potassium and carotene (Table 1.2.1), however, this composition varied according to the variety as reported by Jagadeesh et al. (2007) who observed a wide variation in the TSS, acidity, TSS: acid ratio, sugars, starch and carotenoid contents in the pulp of jackfruit selections of Western Ghats in India.

Glucose, fructose and sucrose are the main sugars of the edible part of both soft and firm varieties of ripe jackfruit, and the variation of carbohydrates and the distribution of free sugar and fatty acid composition of different parts of ripe jackfruit have also been investigated (Chowdhury et al. 1997; Rahman et al. 1999). The composition of jackfruit also varied with the maturity stage and climatic conditions, and both starch and fiber increased with maturity (Rahman et al. 1999).

Table 1.2.1. Nutritional composition of jackfruit pulp (Tee et al. 1997)

Nutrient	per 100g edible portion
Energy	37 kCal
Moisture	83.1 g
Protein	1.6 g
Fat	0.2 g
Carbohydrate	7.3 g
Fibre	5.6 g
Ash	2.2 g
Calcium	37.0 mg
Phosphorus	26.0 mg
Iron	1.7 mg
Sodium	48 mg
Potassium	292 mg
Carotene	110 µg
Vitamin B1 (Thiamine)	66 µg
Vitamin B2 (Riboflavin)	60 µg
Niacin	400 µg
Vitamin C (Ascorbic acid)	7.9 mg

The presence of mannitol in ripe jackfruit was also reported in amounts representing 2% of the firm fruit and 7% of the soft fruit. Ribitol, was also detected in ripe fruit but not in immature fruits (Rahman et al. 1995). The presence of sucrose, glucose and fructose is quite consistent with the findings from other fruits but higher than some other plant material and vegetables (Chowdhury et al. 1997).

On the other hand, jackfruit contains many organic acids predominantly malic acid and citric acid, while succinic acid and oxalic acid were found alongside its esters as the major compound group contributing to its flavour (Ong et al. 2006). The fruit is also a good source of phenolics and flavonoids, compounds that have been found to possess good antioxidant properties (Jagtap et al. 2010b).

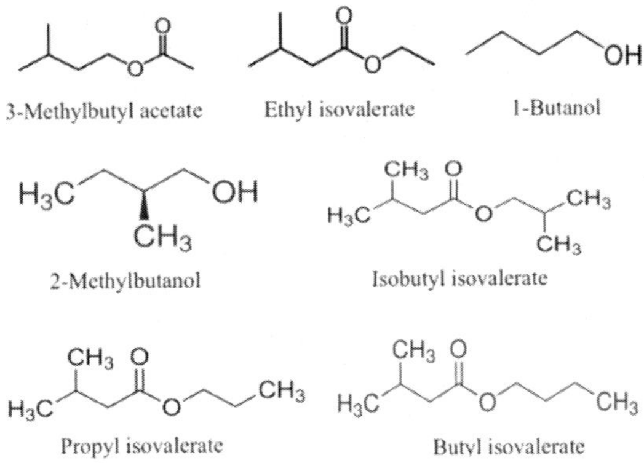

Figure 1.2. The main volatile compounds found in jackfruit (Ong et al. 2008).

Besides these chemical and biochemical compounds, jackfruit also contains many volatile compounds. Maia et al. (2005) investigated the aroma volatiles from two varieties of jackfruit growing in the Amazon using GC–MS, and found that the major components identified in the aroma concentrate of the "hard jackfruit" variety were isopentyl isovalerate and butyl isovalerate, while the aroma concentrate of the "soft jackfruit" was dominated by isopentyl isovalerate, butyl acetate, ethyl isovalerate, butyl isovalerate and 2-methylbutyl acetate. Later and using solid-phase microextraction (SPME) and gas chromatography-time-of-flight mass spectrometry (GC-TOFMS), Ong et al. (2008) identified thirty-seven compounds from five jackfruit cultivars, with the main volatile compounds being: ethyl isovalerate, 3-methylbutyl acetate, 1-butanol, propyl isovalerate, isobutyl isovalerate, 2-methylbutanol, and butyl isovalerate (Figure 1.2), and these compounds contribute to the sweet and fruity note of jackfruit.

2.3. Culinary, Medicinal and Other Uses

2.3.1. Culinary Uses

Jackfruit is considered a staple food in many countries, especially in Asia. Therefore, this crop is subjected to many value-added processes, including canning, freezing and drying of the ripe fruit while the green jackfruit is utilized for the making of preserved foods like pickles, canned and curried vegetables (Morton 1987). In most of Asia, Jackfruit is utilized mainly in the full-grown stage but, is also utilized in the unripe stage. The fruit at this time, may be cut into large chunks for cooking, the only handicap being its gummy latex which accumulates on the knife and the hands unless they are rubbed with oil. The chunks are boiled in lightly salted water until tender, the delicious flesh is cut from the rind and served as a vegetable like a salad rather than a fruit or a dessert. The latex clinging to the pot may be removed by rubbing with oil (Lal et al. 1960). The flesh of the unripe fruit has been experimentally canned in brine or with curry (Bathia et al. 1956). The flesh may also be dehydrated and stored in tins for extensive periods. The tender young fruit may be pickled with or without spices.

The ripe jackfruit on the other hand, produces a very strong odor, which makes it important for its preparation for eating including the extraction of the pulp and seeds from the rind to be conducted outdoors (Popenoe 1974). In the Caribbean and particularly in Jamaica, Antigua and Trinidad and Tobago, the bulbs may be enjoyed raw or cooked (with coconut milk or otherwise); or made into ice cream, chutney, jam, jelly, paste, or canned in syrup made with sugar or honey with citric acid added (Morton 1987). The crisp varieties of jackfruit are preferred for canning. The canned product is more attractive than the fresh pulp, which is sometimes referred to as "vegetable meat" (Morton 1987).

The ripe bulbs can be mechanically pulped to make jackfruit nectar or reduced to concentrate or powder. The addition of synthetic flavoring – ethyl and n-butyl esters of 4-hydroxybutyric acid at 120 ppm and 100 ppm respectively greatly improves the flavor of the canned fruit and the nectar (Morton 1987). In India, the bulbs are dried, fried in oil and salted for eating like potato chips. Canned dried jackfruit pulp is known to have been marketed in Brazil in as early as 1917, and since then many improved methods of preserving jackfruit have been devised, such as frozen ripe bulbs sliced and packed in syrup with citric acid added. The ripe blubs are able to retain a good color, flavor and texture for one year (Morton 1987). The canned product is able to retain quality for up to sixty-three weeks at room temperature, with 3% loss of β-carotene, however when frozen, the canned pulp can be stored for a period of up to two years. The ripe bulbs may also be fermented, distilled and are known to produce potent liquor (Morton 1987).

Jackfruit seeds make up around 10-15% of the total fruit weight (Bobbio et al. 1978). Raw jackfruit seeds are indigestible due to the presence of a powerful trypsin inhibitor. The information on food value per 100 g of edible portion of dried seeds is scanty. The seeds may be boiled and preserved in syrup like chestnuts. They have also been successfully canned in brine, in curry-like baked beans in tomato source and often included in curry dishes. The roasted, dried seeds may be ground to make flour that is blended with wheat flour for baking (Swami et al. 2012). The inedible portion of the jackfruit, the rind, has been found to yield a fair jelly with citric acid. A pectin extract can be made from the peel, undeveloped perianths and core, or just from the inner rind; and this waste also yields syrup used in tobacco curing (Morton 1987). Young jackfruit leaves and young male flower clusters may be cooked and served as vegetables (Morton 1987).

It has also been reported by Hettiaratchi et al. (2011) that the raw jackfruit seeds contain a high amount of resistant starch (indigestible starch). Baking and cooking of the raw seeds can enhance the bioavailability of their nutritional content (Morton 1987). For instance, the flour from jackfruit seeds when prepared was found to be high in protein and carbohydrates. The flour also has good water and oil absorption abilities. Starch was also isolated from the flour that shows high amylose content (Tulyathan et al. 2002).

2.3.2. Medicinal Uses

A number of medicinal properties have been found in jackfruit. For example, the presence of anti-nutritional factors such as tannin and trypsin inhibitor has been reported (Morton 1987), and the Chinese consider jackfruit pulp and seeds to be a tonic that is cooling and nutritious, and was reported to be useful in overcoming the influence of alcohol on the system (Morton 1987). The ash of jackfruit leaves, burned with corn and coconut shells, is used alone or mixed with coconut oil to heal ulcers. Crushed inflorescences are used to reduce bleeding in open wounds (Morton 1987). In Asia, and particularly in India, the starch extracted from seeds is given to relieve biliousness and, trunk wood and the roasted seeds, contain chemical compounds which are considered to have aphrodisiac properties (Le Cointe 1947; Ferrão 1999). The dried latex yields artostenone, a compound with marked androgenic action. Mixed with vinegar, the latex promotes healing of abscesses, snake bite, and glandular swellings.

The roots are used to make a remedy for skin diseases and asthma. An extract of the roots may also be taken to treat fever and diarrhea. The bark is made into poultices. Heated leaves are placed on wounds (Swami et al. 2012). An extract of roots is used in treating skin diseases, asthma and diarrhea. An extract from leaves and latex is used for treating asthma, prevents ringworm infestation, and heals cracking of the feet (Babitha et al. 2004). An infusion of mature leaves and bark is used to treat diabetes and gall stones (Morton 1987). A tea made with dried and powdered leaves is taken to relieve asthma. Heated leaves can treat wounds, abscesses and ear problems, and relieve pain. The wood has a sedative property; its pith is said to precipitate abortion, while the ripe fruit can be used as a laxative (Morton 1987).

2.3.3. Other Uses of Jackfruit

In India and the Philippines, jackfruit is fed to cattle. In some areas, trees are integrated into pastures so that the animals can avail themselves of the fallen fruits. Surplus jackfruit rind is considered a good stock food, whereas the young leaves are eaten by cattle and other livestock and are said to be fattening. In India, the leaves are used as food wraps in cooking, and may be fastened together for use as plates (Haq 2006).

The latex serves as birdlime, alone or mixed with *Ficus* sap and oil from *Schleichera trijuga*. The heated latex is employed as household cement for mending chinaware and earthenware, and to caulk boats and holes in buckets. It has been reported that the chemical constituents of the latex found in jackfruit contains only 0.4 to 0.7% of rubber and cannot be a substitute for it (Mekkriengkrai et al. 2004) however, it contains 82.6% to 86.4%

resins which may have value in varnishes. Its bacteriological activity is equal to that of papaya latex (Mekkriengkrai et al. 2004, Morton 1987).

Timber from jackfruit tree is important in Sri Lanka and to a lesser extent in India as well where it is exported to Europe. It has a rich attractive color which changes with age from orange or yellow to brown or dark-red, it is termite proof, fairly resistant to fungal and bacterial decay, resembles mahogany and is considered to be superior to teak for construction, turnery, masts, oars, implements, brush backs and musical instruments (Best 2015). Palaces were built of jackfruit tree timber in Bali and Macassar, and the limited supply was once reserved for temples in Indochina. Its strength is 75% to 80% that of teak and it also polishes beautifully. Roots of the old tree are greatly prized for carving and picture framing. Dried branches are employed to produce fire by friction in religious ceremonies in Malabar (Best 2015).

From the saw dust or wood chips of jackfruit heartwood, boiled with alum, there is derived a rich yellow die commonly used for dyeing silk and cotton robes of Buddhist priests. In Indonesia, splinters of the wood are put into bamboo tubes collecting coconut toddy in order to impart a yellow tone to the sugar (Indrayan et al. 2004). Besides the yellow colorant, morin, the wood contains the colorless cyanomaclurin and workers in Bombay [Mumbai] reported a new yellow coloring matter, artocarpin, in 1955 (Indrayan et al. 2004). Since then six other flavonoids have been isolated at the National Chemical Laboratory, Poona. The bark also contains 3.3% tannin, which is occasionally made into cordage or cloth (Morton 1987).

3. FRUIT GROWTH AND DEVELOPMENT

The average length of fruit increased rapidly up to 60 days after fruit set and then it decreased again. Fruit diameter increased slowly up to 15 days after fruit set and then the rate of increase was rapid up to 45 days and before slowing down. The development of the circumference followed the same pattern as the length. At maturity, the average length, diameter and circumference attained by the fruit were 35 cm, 24 cm, and 61 cm, respectively (Ullah & Haque 2008). Morton (1987) recorded average length and diameter of jackfruit as 20-90 cm and 15-50 cm, respectively. The changes in weight and volume were slower up to 15 days after fruit set. Thereafter, a rapid increase took place until 105 days. Later, the rate of increase dropped appreciably and continued until maturity (approximately 120 days). At maturity, the average weight and volume attained by the fruits were 6200 g and 6431 cc, respectively.

Length, breadth, and weight of the bulb (pulp + seed) of jackfruit were recorded at 15-day interval after fruit set. They increased rapidly from 15 days after fruit set to 90 days and subsequently slowed down until harvest. Average increase in bulb weight also followed the similar trend as bulb length (Hussain & Haque 1977; Shyamalamma 2007; Ullah & Haque 2008). When harvested at maturity, the average length and breadth attained by the bulb were 5.70 and 3.22 cm whereas the weight was 29 grams. Seed measurements were also recorded at 15 days from fruit set until maturity. The length and breadth of seed increased slowly from 15 to 30 days and then increased rapidly up to 60 days followed by a relatively slow rate of increase until maturity. The rate of increase in seeds' weight was rapid from 15 days up to 60 days from fruit set Hussain and Haque 1977; Ullah & Haque 2008). Then it was slower until maturity. At maturity, the average length and breadth attained by the seeds were 3.9 and 2.2 cm, respectively (Ullah & Haque 2008), while weight of 6g was recorded (Jonathan et al. 2007).

The texture of pulp and seeds also changed with maturity. The pulp texture changed from soft to semi-hard from 15 to 30 days (Table 1.2.2). It turned hard after 45 days and remained so up to 105 days. Again, the bulbs were soft when ripe at 115-120 days. The seeds were soft up to 45 days. Then they became semi-hard up to 75 days and became hard thereafter until they were ripe. From the above results, it is concluded that 45 to 75 days was the optimum time to use jackfruit as a vegetable when the seeds were soft to semi-hard.

Table 1.2.2. Changes in morphological character of fruit, bulb and seed at 15-day intervals from fruit set (Ullah & Haque 2008)

Days after fruit set	Colour Bulb	Colour Seed	Texture Bulb	Texture Seed
15	White	White	Soft	Soft
30	White	White	Semi	Soft
45	White	White	hard	Soft
60	Creamy white	Creamy white	Hard	Semi hard
75	Creamy white	Creamy white	Hard	Semi hard
90	Yellowish white	Light brown	Hard	Hard
105	Yellowish white	Light brown	Hard	Hard
120	Yellowish white Yellow	Brown	Hard Soft	

The growth of jackfruit is characterized by a simple sigmoid curve. This is true when all parameters of measurements length, diameter, circumference, weight, and volume of fruit; length, breadth, weight of bulbs as well as seeds were plotted against time from fruit set to harvesting maturity (Ullah & Haque 2008). A similar growth curve has also been reported for mango, which is botanically a simple fruit but, showed a similar pattern of growth as jackfruit, which is a multiple fruit (Saini et al. 1971).

4. PATHOLOGICAL DISORDERS AND PESTS

Gunasena et al. (1996) reported that the jackfruit is relatively free from serious diseases. This is supported by Soepadmo (1992) who stated that crop protection is not a major concern for growers of jackfruit. This should increase its economic value for growers. Nevertheless, there are some insect pests and pathogens that do affect jackfruit. Principal insect pests in India are the shoot-borer caterpillar, *Diaphania caesalis*; mealybugs. *Nipaecoccus viridis*, *Pseudococcus corymbatus*, and *Ferrisia virgata*, the spittle bug,

Cosmoscarta relata, and jack scale, *Ceroplastes rubina*. The most destructive and widespread bark borers are *Indarbela tetraonis* and *Batocera rufomaculata* (Morton 1987). Other major pests are the stem and fruit borer, *Margaronia caecalis*, and the brown bud-weevil, *Ochyromera artocarpio*. In southern China, the larvae of the longicorn beetles, including *Apriona germarri*, *Pterolophia discalis*, *Xenolea tomenlosa asiatica*, and *Olenecamptus bilobus* cause serious and heavy damage to the fruit stem. The caterpillar of the leaf webbers, *Perina nuda* and *Diaphania bivitralis*, are a minor problem, as are aphids, *Greenidea artocarpi* and *Toxoptera aurantii*; and thrips, *Pseudodendrothrips dwivarna* (Morton 1987; Prakash et al. 2009).

Diseases of importance include pink disease, *Pelliculana* (*Corticium*) *salmonicolor*, stem rot, fruit rot and male inflorescence rot caused by *Rhizopus artocarpi*; and leafspot due to *Phomopsis artocarpina*, *Colletotrichum lagenarium*, *Septoria artocarpi*, and other fungi. Gray blight, *Pestalotia elasticola*, charcoal rot, *Ustilana zonata*, collar rot, *Rosellinia arcuata*, and rust, *Uredo artocarpi*, occur on jackfruit in some regions (Morton 1987).

On the trees, jackfruit may be covered with paper sacks when very young to protect them from pests and diseases similar to what is done with bananas. The bags encourage ants to swarm over the fruit and guard it from its enemies (Morton 1897).

5. Fruit Maturity and Ripening

The jackfruit should be harvested when it has reached an appropriate degree of development and maturity in accordance with characteristics associated with the variety and the area in which they are cultivated. The development and condition of the jackfruit at the time of harvest must be such as to enable it to withstand handling, packing and transportation (Anonymous 2009).

The fruit ripens normally at tropical ambient temperatures (20-35°C) in three to ten days depending on the stage of maturity at harvest and with no problems to delay ripening. Starch is the principal storage material in the bulb and during ripening it is converted to sugars. The color of the bulbs changes from pale to light yellow to an attractive golden yellow color and is accompanied by the characteristic, sweet aroma (Anonymous 2009). Fruit growth and maturation normally takes 5 months after fruit set but harvesting can be done even after 4 months. In cooler places and higher altitudes fruit maturation takes longer. Three stages of maturation are distinguished: immature; mature for cooking; and ripe for eating fresh. Immature fruits- those that are dark green with closely spaced spines- are usually cooked as vegetable (Anonymous 2009).

There are several characteristics of the fruit that can be used as indicators of maturity, either on their own or together for a particular cultivar (Palang & Cajes 2000). The best indicator may be a solid sound when tapped. This indicates readiness for harvesting as green fruit whereas ripe fruit has a hollow sound. In many cultivars, the skin color changes from green to light green or yellow and gives off a strong aroma, and the spines on the skin become flattened and wider.

Yap (1972) described the following characteristics of fruit maturity:

- Rind changes color from green to yellow or greenish yellow.
- Spines become well-developed and well-spread and yield to moderate pressure.
- Last leaf on the footstalk turns yellow.
- Fruit produces a dull hollow sound when tapped by a finger.

Among these indicators of maturity, the last is claimed to be the most reliable. The other indicators, particularly the yellowing of the last leaf of the footstalks, appear to be inadequate as sole indices. Angeles (1983) observed that some footstalks with no remaining leaves still had developing fruit. Likewise, the change in rind color as a basis for determining fruit maturity is not a dependable indicator. Some cultivars turn light green, greenish-yellow, yellow, yellowish-brown or rusty brown at the mature stage.

6. POSTHARVEST HANDLING

6.1. Harvesting

Jackfruit is harvested at different stages of maturity depending on the intended use. Jackfruit matures 3-8 months after flowering. Since the ripening proceeds normally even when the fruit is harvested a little earlier than at the optimum mature stage, it is not essential to wait until full maturity to harvest it (Haq 2006). However, there are some differences in the taste from fruit to fruit, probably due to variation in maturity. Larger fruits give bulbs that are tastier than smaller ones (Haq 2006). Young fruits which are green to yellowish in color and with well expanded spines, are harvested for use as vegetables during the first two to three months after fruit-set, before the seeds harden, and mature fruits are harvested after four months for dessert purposes. However, the timing depends on the date of emergence of the inflorescence (Haq 2006).

Harvesting is carried out by cutting the peduncle with a sharp knife and by traditional methods, such as the use of ropes and sickles for upper fruit harvesting and hand picking for lower fruits (Haq 2006). If fruits fall to the ground, they will bruise and deteriorate, unless consumed immediately. They should be harvested into a sack or lowered by a rope to avoid damage. If not harvested at maturity the fruits ripen further on the tree giving off its strong aroma (Haq 2006). The fruit should be consumed soon after harvest (Teaotia & Awasthi 1968). Deformed fruit is harvested early because the flesh will not develop well. Harvesting of the fruits as soon as they reach maturity facilitates better handling and transport. For the processing of jackfruit, a uniform maturity index of jackfruit is needed to obtain a quality-processed product (Haq 2006; Teaotia & Awasthi 1968).

Jackfruit taste is substandard if harvested from young plants. Old trees are preferred since sweetness and taste increasingly improve with the advancing age of the tree. Fruit growth and maturation normally takes five months after fruit set, but harvesting of immature fruits can take place as early as four months (Teaotia & Awasthi 1968). In cooler places and higher altitudes, fruit maturation takes longer. For culinary purposes, tender jackfruit is harvested within two to three months after fruit set, before the seeds harden. Under northern Indian conditions, the optimum maturity for harvesting good quality ripe fruit is about 180 days from the date of spike emergence (Teaotia & Awasthi 1968).

6.2. Handling

In South Asian countries like India, Bangladesh and Sri-Lanka, harvested fruits are carried individually by holding the stalk, and are generally loaded into bullock-carts or pushcarts and transported to nearby towns or village markets for retail sales, or are sold wholesale to visiting tradesmen from larger towns (AFPAD n.d). However, during the peak production period, trucks are used to transport them to markets. The fruit is sometimes transported over 1000 km in order to attain better market prices. Generally, no packaging material is used for transportation. Fruit is carried manually in baskets or bags from the field up to the road point. Post-harvest losses can be as much as 30-34% (ICUC 2005).

There are several steps that can be taken to improve the quality of jackfruit and reduce post-harvest losses (Kader 2009). The first of these is careful handling and packaging to avoid bruising during transport and storage. For efficient marketing and utilization, harvested fruit should be stored according to the stage of maturity and ripeness. This allows the most mature and ripest fruits to be sold first (Haq 2006). The less ripe fruits can be put aside and allowed to further ripen. For marketing, the fruit should also be graded according to size-large fruits weighing over 16 kg, medium fruits weighing between 8-16 kg and small fruits below 8 kg (Haq 2006). Prior to use, fruits should be washed to remove latex stains and dust. It is not appropriate or necessary to wash the fruits prior to transportation (ICUC 2005).

ICUC (2005) recommended that post-harvest operations of jackfruit should be practiced in seven steps: (i) post-harvest operation, (ii) packaging and storage of fresh fruits, (iii) ripening, (iv) pre-processing into fruitlets, (v) packaging and storage of jackfruit fruitlets, (vi) pre- processing into pulp, and (vii) packaging and storage of pulp. There is a need to wash fruits before processing to remove dirt, latex stains and any field contamination and drain them properly to remove excess moisture from the surface of the fruit.

6.3. Storage

As mentioned earlier, jackfruit has a storage life of three to ten days depending on the maturity, ambient temperature and relative humidity (RH) conditions, although the fruit is not normally stored in cold storage. Mathur et al. (1952) studied a range of temperatures from 0-28°C and relative humidity (RH) ranging from 55-90% and reported that 11-13°C 85-90% RH is the optimum for storing jackfruit. Singh (2000) also reported a storage life of six weeks at 11-12.7°C and 86-90% RH.

During storage, the sucrose content decreases from an initial value of 9.5% to around 5% and is accompanied by a rise in reducing sugars from 2% to 6%. However, ascorbic acid reduction also occurred from an initial 8.2 to 3.5% (Mathur et al. 1952).

At the farmers' level in Bangladesh, harvested mature fruits are stored in the corner of houses to ripen for one week. Some farmers insert common salt into the floral stalk for quicker ripening. Sometimes farmers dig holes in the ground and store unripe jackfruit (Haq 2006).

7. Processing

Post-harvest loss is high due mainly to spoilage. To overcome this, producers generally sell their marketable surplus within a short time from harvest at low prices. However, income can be significantly increased, if produce is stored correctly or processed, since fruit prices can double or even triple only a few months after the harvest (Roy 2000).

Oliveros et al. (1971) reported that the demand for ripe jackfruit increased 100-fold in the Philippines following modern advances in food technology. Reduced post-harvest losses, increased shelf-life and preserved fruit for the out of season period, can improve the use of jackfruit through processing. Raw materials transformed into edible products can increase food security and add variety to peoples' diet thus improving nutrition and health. Creation of income and employment opportunities in production areas is an added bonus. The bulbs possess a desirable texture and a rich appetizing taste. In shredded form, they may be eaten raw or used as an ingredient in ice cream, candies and other forms of desserts (Swami et al. 2012).

More recently, consumer satisfaction with minimally processed jackfruit, such as ready-to-eat jackfruit slices, was tested as the demand for this produce in its various forms is increasing. Modified atmosphere packaging (MAP) of fresh cut jackfruit was utilized to extend the shelf-life of the bulbs. Results showed that MAP with low oxygen and higher carbon dioxide levels, fresh-cut jackfruit shelf-life was extended significantly when stored at refrigerated temperature (Saxena et al. 2008, Sudiari 1997). MAP was also found to not only preserve the shelf-life of fresh cut jackfruit, but the phytochemical compounds were also well preserved during storage of jackfruit bulbs under different gases composition (Saxena et al. 2009; Sudiari 1997). Recently, interesting work was also conducted by Vargas-Torres et al. (2017) by investigating the effect of pre-treatment with 1-methylcyclopropene (1-MCP) and edible coatings quality and shelf life of pre-cut jackfruit. When stored at 5 °C and coated with 1-MCP/edible coatings, good quality attributes and sensory acceptability of the fresh-cut jackfruit were maintained with longer shelf-life.

Oliveros et al. (1971) extracted an essential oil from the rind by steam distillation. However, the yield was low (0.03%). It was colorless with a refractive index (np) 25 = 1.48 and density (d) 25 = 0.912. They concluded that for profitable commercial extraction of the essential oil more work was needed in the food and perfumery industries. Abraham (1971) suggested that steam distillation destroys the aroma and possibly is not the best method for extracting the flavor components of jackfruit. Nataranjan and Karunanithy (1974) have shown that the natural aroma could be added back in preserved products, i.e., canned products and nectars, by the addition of a few milligrams per liter of two esters synthesized from gamma-butyrolactone.

Swords et al. (1978) and later Rasmussen (1983) identified 20 major flavor components of jackfruit.

Singh and Mathur (1954) investigated the freezing of jackfruit bulbs. The edible bulbs from ripe fruits (excluding the seeds) were sliced and packed (i) with dry sugar, and (ii) in 50% sugar syrup with 0.5% citric acid (on the basis of syrup), into jam cans. The product was frozen at -29°C and subsequently stored at -18°C. Slices packed in 50% sugar syrup containing 0.5% citric acid gave excellent results. The color, taste and flavor of the product were preserved for one year. Moreover, the individuality of the slices was maintained for the year. The product prepared with dry sugar had a lower quality.

More recently, jackfruit has been subjected to some agro-processing. A number of products have been developed from raw, tender and ripe fruits and seeds. There are at least five products prepared from tender fruits. They are (i) canned in brine, (ii) canned in curried form (in combination with other vegetables), (iii) made into dehydrated products, (iv) papadam or papad, (v) sweet or sour pickles in oil and (vi) sweet and sour pickle in vinegar (Bhattacharjee 1981).

The ripe fruit bulbs (excluding seed) and the rind of the ripe fruit (including perianth and unfertilised flowers) have also been used for processing in a number of products. Ripe jackfruit bulbs are canned in syrup, made into jams either pure or mixed with dehydrated bulbs, chutney, preserves, candy, and concentrate and powder. Rinds of ripe fruit are made into syrup, pectin, jelly, pectin extracts, biscuits and papadam (Bhattacharjee 1981). The seeds of ripe fruit are used as vegetables when cooked with grated coconut, chili, salt and spices and are a popular side dish with rice (Thomas 1980). Roasted or fried seeds are tasty to eat when stewed with meat. Seeds can be processed: (i) canned in brine, (ii) canned in a curried style such as in combination with other vegetables, (iii) canned in tomato sauce, (iv) dried to make flour, and (v) roasted.

CONCLUSION

Jackfruit is indeed a multifaceted fruit gaining popularity worldwide. This exotic commodity has considerable potential to generate income in particular in rural areas and alleviate malnutrition in impoverished communities. The products and by-products that can be produced from this fruit can be utilized in many aspects of our everyday life. However, due to its high rate of perishability more needs to be understood about the physiological and biochemical modifications that occur during pre-harvest and postharvest stages of ripening so we can understand better how to preserve the quality, taste, flavor and extend the shelf live so a wider audience can have access to the fruit.

Jackfruit has great commercial and food and nutrition security potential but these are largely untapped right now. This is because many people are unaware of the diverse ways the product can be used for food, medicine, and construction. In many places where food insecurity is a concern it is used only as fresh fruit. Public education campaigns should be used to sensitize the public to its potential. Integrating this information in public school curricula, will also help to disseminate information about the varied uses of jackfruit.

REFERENCES

Abraham, KO, 1971, *'Fruit flavors*: Preliminary studies on jackfruit *(Artocarpus integrifolia L.) flavour'* MSc. thesis, University of Mysore.

Agricultural and Processed Food Products Export Development Authority n.d., *Comprehensive master plan for tapping the export potential of North Eastern states*, APEDA, Hyderabad, viewed 23 August 2017. http://apeda.gov.in/apedawebsite/Announcements/APEDA_NER_Final_Project_Report_Compr.pdf.

Ahmed, K, Malek, M, Jahan, K, & Salmatullah, K, (eds.) 1986, *Nutritive Value of Food Stuff*, 3rd edn, University of Dhaka/Institute of Nutrition and Food Science, Bangladesh.

Alexander, DMP, Scholefield, PB, & Frodsham, A 1983, 'Jackfruit. In: *Some tree fruit from tropical Australia*. Report # 27. Commonwealth Scientific and Industrial Research Organisation, Canberra.

Angeles, DO, 1983., 'Cashew and jackfruit research', Research Technical Report, Department of Horticulture, University of Philippines at Los Banos (UPLB), Laguna.

Anonymous 2009, *'Proposed Draft Brunei Standard for Jackfruit,'* Department of Agriculture, Ministry of Industry and Primary Resources. Brunei Darussalam.

Babitha, S, Sandhya, C, & Pandey, A, 2004, 'Natural food colorants,' *Applied Botany Abstracts*, vol. 23, pp. 258-266.

Best, A, 2015, 'Value-adding options for tropical fruit using jackfruit as a case study,' *Rural Industries Research and Development Cooperation, Publication No 15/042*, Barlton, Australia.

Bhattacharjee, S, 1981, *'Chemistry and technology of some minor fruits'* MSc. thesis, University of Mysore.

Bobbio, FO, El-Dash, AA, Bobbio, PA, & Rodrigues, LR, 1978 'Isolation and characterization of the physicochemical properties of starch of jackfruit seeds (*Artocarpus heterophyllus*),' *Cereal Chemistry*, vol. 55, pp. 505–511.

Chowdhury, FA, Raman, MA, & Mian, AJ, 1997, 'Distribution of free sugars and fatty acids in jackfruit (*Artocarpus heterophyllus*),' *Food Chemistry*, vol. 60, pp. 25–28.

Chaudhary, UL & Khatari, BB, 1997, *'A status report on jackfruit, pummelo and mangosteen'*, UTFANET, Southampton University.

Corner, EJH, 1938, 'Notes on the systematics and distribution of Malayan phanerogams II. The jack and the chempedak', *Gardener Bulletin*, vol. 10, pp. 56-81.

Elevitch, CE & Manner, HI, 2006, *Artocarpus heterophyllus* (jackfruit). *Species profiles for pacific island agroforestry*, viewed 20 September 2017, http://www.agroforestry.net/tti/A.heterophyllus-jackfruit.pdf.

Ferrao, JEM, 1999, *Fruticultura tropical: especies com frutos comestiveis*, vol. I. Instituto de Investigacao Cientifica Tropical, Lisboa.

Gunasena, HPM, Ariyadasa, KP, Wikramasinghe, A, Herath, HMW, Wikramasinghe, P, & Rajakaruna, SB, 1996, '*Manual of jackfruit cultivation in Sri Lanka,*' Manual # 48. Forest Information Service, Forest Department, Sri lanka.

Haq, N 2006, '*Jackfruit, Artocarpus heterophyllus*', Southampton Centre for Underutilised Crops, University of Southampton.

Hettiaratchi, UPK, Ekanayake, S, & Welihinda, J, 2011, 'Nutritional assessment of a jackfruit (*Artocarpus heterophyllus*) meal,' *Ceylon Medical Journal*, vol. 56, pp. 54–58.

Hussain, M & Haque, A, 1977, 'Studies on the physical characteristics of jackfruit,' *Bangladesh Horticulture*, vol. 5, pp. 9–14.

Indrayan, AK, Kumar, R, Rathi, AK, 2004, 'Multibeneficial natural material: dye from heartwood of *Arthocarpus heterophyllus* Lamk,' *Journal of Indian Chemical Society*, vol. 81, pp. 1097–1101.

International Centre for Underutilised Crops 2005, 'Training manual on processing and small business development,' *ICUC*, University of Southampton.

Jacob, P & Narasimhan, P, 1992, 'Processing and evaluation of carbonated beverage from jackfruit waste,' *Journal of Food Processing Preservation*, vol. 16, pp. 373–380.

Jagadeesh, SL, Reddy, BS, Basavaraj, N, Swamy, GSK, Gorbal, K, Hegde, L, Raghavan, GSV, & Kajjidoni, ST, 2007, 'Inter tree variability for fruit quality in jackfruit selections of Western Ghats of India,' *Scientia Horticulturae*, vol. 112, pp. 382–387.

Jagtap, UB. Panaskar, SN, & Bapat, VA, 2010, 'Evaluation of antioxidant capacity and phenol content in jackfruit (*Artocarpus heterophyllus* Lam.) fruit pulp,' *Plant Foods for Human Nutrition*, vol. 65, pp. 99–104.

Jagtap, UB & Bapat, VA, 2010, '*Artocarpus*: A review of its traditional uses, phytochemistry and pharmacology,' *Journal of Ethnopharmacology*, vol. 129, pp. 142–166.

Jonathan, HC., Balerdi, CF, & Maguire, I, 2007, *'Jackfruit growing in the Florida home landscape,* viewed 20 September 2017, http:/Jedis.ifas.ufl.edu/MG370.

Kader, AA 2009, *'Jackfruit: Recommendations for maintaining postharvest quality,'* Department of Plant Sciences, University of California at Davis, viewed 23 October 2017, http://Postharvest.ucdavis.edu/Produce/ProduceFacts/Fruit/jackfruit.shtml.

Lal, G, Siddappa, GS, & Tandon, GL, 1960, *Preservation of fruits and vegetables*, Indian of Agricultural Research, New Delhi.

Le Cointe, P, 1947, *Amazonia Brasileira III. Arvires e plantas uteis.* [Brazilian Amazonia III. Useful Trees and plants.] Companhia editora Nacional, Rio de Janeiro.

Maia, JGS, Andrade, EHA, & Zoghbi, MGB, 2004, 'Aroma volatiles from two fruit varieties of jackfruit (*Artocarpus hetero*phyllus Lam.),' *Food Chemistry*, vol. 85, pp. 195–197.

Manalo, MLA, 1986, *'The relationship of growth and flowering habits of jackfruit (Artocarpus heterophyllus Lam.'* BSc. thesis, University of Philippines at Los Banos, Laguna.

Mathur, PB, Singh, KK, & Kapur, NS, 1952, 'A note on investigations on the cold storage and freezing of jackfruit,' *Indian Journal of Horticulture*, vol. 11, pp. 149-153.

Mekkriengkrai, D, Ute, K, Swiezewska, E, Chojnacki, T, Tanaka, Y, & Sakdapipanich, JT, 2004, 'Structural characterization of rubber from jackfruit and euphorbia as a model of natural rubber', *Biomacromolecules*, vol. 5, pp. 2013–2019.

Moncur, MW, 1985, *'Floral ontogeny of the jackfruit, Artocarpus heterophyllus Lam. (Moraceae)'*, Division of Water and Land Resources, CSIRO, Canberra.

Mondal, C, Remme, RN, Mamun, AA, Sultana, S, Ali, MH & Mannan, MA, 2013, 'Product development from jackfruit (*Artocarpus heterophyllus*) and analysis of nutritional quality of the processed products', *IOSR Journal of Agriculture and Veterinary Science*, vol. 4, pp. 76–84.

Morton, JF, 1987, *'Fruits of warm climates,'* Creative Resources System, Winterville, North Carolina.

Munishamanna, KB, Ranganna, B, SubramanyaS, Chandu, R, & Palanimuthu, V, 2007, 'Development of value-added products from jackfruit (Artocarpus heterophyllus L.) to enhance farm income of rural people," in PG Chengappa, N Nagaraj & R Kanwar (eds.), *International Conference on 21st Century Challenges to Sustainable Agri-Food Systems: Biotechnology, environment, nutrition, trade and policy*, I. K. International Publishing House, New Delhi.

Narasimham, P, 1990, 'Breadfruit and jackfruit,' in S Nagy, PE Shaw, & WF Wardowski (eds.), *Fruits of tropical and subtropical origin: composition, properties and uses*, Florida Science Source Inc., Florida.

Nataranjan, PN & Karunanithy, R, 1974, 'Synthetic flavour enhancers for *Artocarpus integrifolia*,' *The Flavour Industry*, vol. 5, pp. 282–283.

Oliveros, BL, Cardeno, V, & Perez, P, 1971, 'Physical properties of some Philippine essential oils', *The Flavour Industry*, vol. 2, pp. 305–309.

Ong, BT, Nazimah, SAH, Osman, A, Quek, SY, Voon, YY, Hashim, DM, Chew, PM, & Kong, YW, 2006, 'Chemical and flavour changes in jackfruit (*Artocarpus heterophyllus* Lam.) cultivar J3 during ripening,' *Postharvest Biology and Technology*, vol. 40, pp. 279–286.

Ong, BT, Nazimah, SAH, Tan, CP, Mirhosseini, H, Osman, A, Mat Hashim, D, & Rusul, G, 2008, 'Analysis of volatile compounds in five jackfruit (*Artocarpus heterophyllus* L.) cultivars using solid-phase microextraction (SPME) and gas chromatography-time-of-flight mass spectrometry (GC-TOFMS),' *Journal of Food Composition and Analysis*, vol. 21, pp. 416–422.

Palang, DB & Cajes, BC, 2000, *Maturity indices of two types jackfruit*, Agency In-house Research & Review Report. EVIARC Balinsasayao, Abuyog, Leyte, Philippines.

Popenoe, W, 1974, *Manual of tropical and subtropical fruits*, Hafner Press, Macmillan Publishing, New York, NY.

Prakash, O, Kumar, R, Mishra, A, & Gupta, R, 2009, '*Artocarpus heterophyllus* (Jackfruit): an overview,' *Pharmacognosy Reviews*, vol. 3, pp. 353-358.

Punan, MS, Abdulah Rahman, AS, Nor, LM, Muda, P, Sapii, AT, Yon, RM, & Som, FM, 2000, 'Establishment of a quality assurance system for minimally processed jackfruit,' in JL Van To, N Duy-Duc, & MC Webb (eds.), *Quality assurance in agricultural produce, ACIAR Proceedings*, vol. 100, pp. 115-122.

Purseglove, JW, 1974, *Tropical crops: Dicotyledons*, vol. 2, Longman Scientific and Technical, John Wiley & Sons, New York, NY.

Rahman, AKMM, Huq E, Mian, AJ, & Chesson, A, 1995, 'Microscopic and chemical changes occurring during the ripening of two forms of jackfruit (*Artocarpus heterophyllus* L.),' *Food Chemistry*, vol. 52, pp. 405–410.

Rahman, MA, Nahar, N, Mian, AJ, & Mosihuzzaman, M, 1999, 'Variation of carbohydrate composition of two forms of fruit from jack tree (*Artocarpus heterophyllus* L.) with maturity and climatic conditions,' *Food Chemistry*, vol. 65, pp. 91–97.

Rasmussen, P, 1983, 'Identification of volatile components of jackfruit by gas chromatography/mass spectrometry with two different columns,' *Analytical Chemistry*, vol. 55, pp. 1331–1335.

Roy, SK, 2000, Promotion of underutilised tropical fruit processing and its impact on world trade,' *Acta Horticulturae*, vol. 518, pp. 233–236.

Saini, SS, Singh, RN, & Paliwal, GS, 1971, 'Studies on fruit set, growth and development in mango (*Mangfera indica* L.),' *Indian Journal of Horticulture*, vol. 28, pp. 247-256.

Saxena, A, Bawa, AS, & Raju, PS, 2008, 'Use of modified atmosphere packaging to extend shelf-life of minimally processed jackfruit (*Artocarpus heterophyllus* L.) bulbs,' *Journal of Food Engineering*, vol. 87, pp. 455-466.

Saxena, A, Bawa, AS, & Raju, PS, 2009, 'Phytochemical changes in fresh-cut jackfruit (*Artocarpus heterophyllus* L.) bulbs during modified atmosphere storage,' *Food Chemistry*, vol. 115, pp. 1443-1449.

Shyamalamma, S, 2007, *Study of genetic diversity in jack (Artocarpus heterophyllus Lam.) using morphological and molecular markers*, Ph. D. thesis, University of Agricultural Sciences, Bangalore.

Singh, RP, 2000, 'Scientific principles of shelf-life evaluation. In: CMD Man & AA Jones (eds.), *Shelf-life evaluation of foods*, 2nd edn. Aspen Publishers, Maryland, MD.

Singh, KK & Mathur, PB, 1954, 'A note on investigations on the cold storage and freezing of jackfruit,' *Indian Journal of Horticulture*, vol. 11, pp. 149-154.

Soepadmo, E, 1991, '*Artocarpus heterophyllus* Lam,' in EWM Verheij & RE Coronel (eds.). *Plant resources in South East Asia*, vol. 2: Edible fruits and nuts, Pudoc, Wageningen.

Sudiari, NM, 1997, '*Studies on the storage characteristics of minimally processed jackfruit stored under modified atmosphere packaging (Artocarp s heterophyllus Lam.)*,' B. Sc. Thesis, Bogor Agricultural University, Bogor.

Swami, SB, Thakor, NJ, Haldankar, PM, & Kalse, SB, 2012, 'Jackfruit and its many functional components as related to human health: A review,' *Comprehensive Reviews in food Science and Food Safely*, vol. 11, pp. 565–576.

Swords, G, Bobbio, PA, & Hunter, GLK, 1978, 'Volatile constitution of jackfruit *Artocarpus heterophyllus*,' *Journal of Food Science*, vol. 43, pp. 639–640.

Teaotia, SSK & Awasthi, RK, 1968, 'Dehydration studies in jackfruit (*Artocarpus heterophyllus* Lam),' *Indian Food Packer*, vol. 22, pp. 6–14.

Tee, ES, Mohd Ismail, N, Mohd Nasir, A, & Khatijah, I, 1997, '*Nutrient composition of Malaysian foods,*' Institute Medical for Research, Kuala Lumpur.

Thomas, CA, 1980, 'Jackfruit, *Artocarpus heterophyllus* Lam. (Moraceae) as source of food and income,' *Economic Botany*, vol. 3, pp. 154–159.

Tulyathan, V, Tananuwong, K, Songjinda, P, & Jaiboon, N, 2002, 'Some physicochemical properties of jackfruit (*Artocarpus heterophyllus* Lam) seed flour and starch,' *Science Asia*, vol. 28, pp. 37–41.

Ullah, MA & Haque, MA, 2008, 'Studies on fruiting, bearing habit and fruit growth of Jackfruit germplasm,' *Bangladesh Journal of Agricultural Research*, vol. 33, pp. 391–397.

Vargas-Torres, A, Becerra-Loza, AS, Sayago-Ayerdi, SG, Palma-Rodríguez, HM, García-Magañab, ML, & Montalvo-González, F, 2017, 'Combined effect of the application of 1-MCP and different edible coatings on the fruit quality of jackfruit bulbs (*Artocarpus heterophyllus* Lam) during cold storage,' *Scientia Horticulturae*, vol. 214, pp. 221-227.

Yap, AR, 1972, *'Cultural directions for Philippine agricultural crops: Fruits,'* Department of Agriculture, Manilla.

In: Agriculture, Food, and Food Security
Editor: Clinton Lloyd Beckford

ISBN: 978-1-53613-483-4
© 2018 Nova Science Publishers, Inc.

Chapter 2

INFLUENCING FACTORS ON THE IODINE CONTENT OF FOOD: AN INTRODUCTORY REVIEW OF NON-ANIMAL I-SOURCES

Gerhard Flachowsky[*]
Institute of Animal Nutrition, Friedrich-Loeffler-Institute
Braunschweig, Germany

1. INTRODUCTION

Iodine (I) is an essential trace element for humans and animals. It was discovered as a novel element about 200 years ago (see Küpper et al. 2011) and two centuries of iodine research have shown many interesting results, but also opened some new questions in animal and human nutrition (European Food Safety Authority (EFSA) 2005; EFSA 2013a; EFSA 2013b; EFSA 2014; Küpper et al. 2011). Iodine is part of the thyroid hormones triiodothyronine (T3) and thyroxine (T4) and therefore, it is highly important for key processes in the body (see (Decuypere et al. 2005; EFSA 2014;

---

[*] Corresponding author email: Gerhard.Flachowsky@fli.de.

Zimmermann & Andersson 2012). For example, a lack of iodine is known to cause a loss of viability in embryos. Furthermore, the development of the human brain may be adversely affected. A severe iodine deficiency during pregnancy is known to cause mental development delays up to cretinism. Goiter- the enlargement of the thyroid gland- is another clinical sign of an insufficient supply of iodine. Further health consequences may be hypothyroidism and iodine induced hyperthyroidism.

Nutritional, biochemical, pathological, and therapeutical aspects of iodine in humans and animals were recently summarized by some authors (see EFSA 2014; McDowell 2003; National Research Council (NRC) 2005; Preedy, Burrow & Watson 2009; Suttle 2010). Iodine deficiency still remains a major public health issue in many countries, including some European countries (Zimmermann & Andersson 2012). In 2001, the World Health Organization (WHO) estimated that about 800 million people would suffer from iodine undersupply- defined as an Iodine concentration less than 100 μg /L urine (EFSA 2014; WHO 2004; WHO, *International Council for Control of Iodine Deficiency Disorders* (ICCIDD) & United Nations Children's Fund (UNICEF) 2001; WHO, ICCIDD & UNICEF 2007)- but altogether the number of countries with insufficient supply of iodine is decreasing - from 54 in 2003 to 30 in 2012 (Andersson, Karumbunathan & Zimmermann 2012; Pearce, Andersson & Zimmermann 2013; Zimmermann & Andersson 2012). On the other hand, there are ten countries where people showed excessive iodine intake (Pearce, Andersson & Zimmermann 2013).

Extensive efforts to improve iodine supply in human nutrition, for example by supplementing iodine to table salt by the food industry and in households (Andersson, de Benoist & Rogers 2010; van der Haar et al. 2011; Zimmermann & Andersson 2011) or by enriching foods of animal origin by iodine supplementation in feeds extending animal requirements (Flachowsky et al. 2014; Schöne & Rajendram 2009) contributed to a better iodine supply.

In some countries 40% and more of the total human iodine intake derives from milk and milk products (Chacón Villanueva & Agency 2016; Dahl et al. 2003; Johner et al. 2012). Besides milk, eggs of laying hens are also characterized by a high transfer of feed-iodine into the animal product, especially into the egg yolk (Opalinski 2017; Röttger 2012; Slupczynska et al. 2014) and may contribute to 10% and more of the total human iodine intake (Chacón Villanueva & Agency 2016; Sager 2011). Recently, Hester (2017) considered the innovation potential of eggs for improvement of the human I-supply. The iodine intake from drinking water, tea, coffee and other beverages has also been the subject of research and has been found to be dependent on the I-content of water which may vary between below 1 and up to 25% - for example in Denmark (Andersen et al. 1996; Rasmussen, Larsen & Ovesen 2000) and more.

It should be mentioned, that there is just a small range between the average iodine requirements of humans which is 130 – 200 µg I per day for adults (EFSA 2014) (see Table 1) and the upper level - UL - of 500 - 600 µg I per day for adults (Deutsche Gesellschaft für Ernährung, Österreichische Gesellschaft für Ernährung, Schweizerische Gesellschaft für Ernährungsforschung, Schweizerische Vereinigung für Ernährung (DACH) 2000; Scientific Committee on Food (SCF) 2002), so that the maximum level is only about three times higher than average human requirement.

Thus, on the one hand there is a certain risk of deficiency, but on the other hand also a risk of overconsumption, especially if people consume high amounts of iodine-fortified food of animal origin such as milk and eggs, as well as high amounts of iodized salt (WHO, ICCIDD & UNICEF 2007) or when they are living in regions with high iodine concentrations in drinking water (Li et al. 1987; Liu et al. 2009; Lu et al. 2005; Shen et al. 2011; Zimmermann 2009) Therefore, dose-response studies with animals, characterized by a high carry-over of iodine from feed into their food products, were required by the EFSA (EFSA 2005).

The most important objectives of such studies can be summarized as follows:

1. A better evaluation of the contribution to human supply made by animal food-products.

2. Knowledge of further iodine inputs and the effects of antagonists on the iodine content of food of animal origin.
3. Influence of further factors, such as animal species, categories and breeds, animal keeping, hygiene measurements etc. on iodine-content of food of animal origin.

For preventive consumer protection, the EFSA in 2005 recommended a reduction of the iodine-upper level in feed of dairy cows and laying hens from 10 to 5 mg I per kg feed with 88% dry matter (DM). Later, EFSA proposed a further reduction of iodine in feed of dairy cows (2 mg/kg) and laying hens (3 mg/kg mixed feed) (EFSA 2013a; EFSA 2013b).

There are many sources of iodine for human consumption. These include major sources like animals, poultry, plant, water, and salt. In this chapter, the focus is on non-meat I-sources. Specifically, the chapter discusses vital aspects of various plant food sources- fruit, vegetables, and seaweeds- water, and iodized salt as influencing factors in human iodine intake. Before this discussion, some foundational principles regarding iodine intake requirements and recommendations for human health, are explored.

2. Requirements and Excess of Iodine in Human Nutrition

In various regions of the world many people - about 800 million - still today suffer from iodine deficits (de Benoist et al. 2008; Zimmermann & Andersson 2012). The iodine requirements, or the recommended iodine intake for humans, are given as a function of age, body weight, physiological stage, and gender and are shown in Table 2.2.1. They vary depending on the influencing factors between 40 (0 – 1 year old) and 290 µg/day (during lactation).

Table 2.2.1. Recommended iodine intake of humans (µg/day) by various scientific bodies

Age/ Physio- logical stage	DACH (2000)	WHO, ICCIDD & UNICEF (2001)	Institute of Medicine (IOM) (2001)	Australian Government (2006)	Nordic Nutrition Recommen- dations (NNR) (2012)	EFSA (2014)
0-1 year	40-80		110-130	90 - 110		70
0-6 years		90			90	
1-8 years			90	90		
1-10 years						90
1-15 years	100-200					
6-9 years					150	
6-12 years		120				
9-13 years			120		120	
11-14 years				120		120
14-18 years/ Adults	180-200	150	150		150	
15-17 years				150		130
>18 years				150		150
Pregnancy	200-230		220	220	175	200
Lactation	260		290	270	200	200
Pregnancy/ Lactation		200				

In general, there are no large differences between the iodine requirements/recommendations of various societies (Table 2.2.1). The Nordic Nutrition Recommendations (NNR) (NordicNutrition Recommendations 2012) distinguished average requirements (AR) from the recommended average intake (AI). The AR is estimated to be 100 µg/d for adults, but the recommended intake is set to 150 µg/day (Table 1). Andersson et al. (2007) proposed to increase the iodine intake during pregnancy from 200 to 250 µg/day.

In human nutrition, the iodine upper levels (UL) vary between 200 (1 to 3 years old babies and infants) and 1 100 µg per day (adults) (Australian

Government 2006; Institute of Medicine (IOM) & (US) Panel on Micronutrients 2001). The Joint FAO/WHO-Expert Committee on Food Additives (JECFA) set a tolerable daily intake for iodine from all sources of 1 mg/day (17 µg/kg body weight) (JECFA 1990). In Europe, the DACH (2000) proposed a limit of 500 and the SCF (2002) of 600 µg I/day for adults. The margin between demand level and maximum level for iodine in humans is slight- demand for adults 200 µg UL in Europe, that is 500-600 µg/day with a ratio of: 1: 2.5-3- so that the maximum level is only 2.5 to 3 times higher than human demand (for more details see EFSA 2014). Gunnarsdottir and Dahl (2012) developed a systematic literature review about iodine intake in human nutrition and came to the conclusion that there were no new data supporting the need for changes in the dietary reference values for children or adults.

Based on seaweed consumption, much higher I-intakes are reported for Japan and other South-East Asia countries. Zava and Zava (2011) reviewed studies based on urine iodine analysis that showed daily intakes between 1 and 3 mg I. Also, Teas et al. (2004) mentioned that Asian seaweed dishes may exceed the tolerable upper iodine intake level of 1 100 µg/day.

3. Iodine Requirements and Excess of Food Producing Animals

The iodine requirements of animals depend on the species, the thyroid activity and the metabolic rate. That means that the requirements are remarkably similar when expressed in terms of energy intake and heat production (Suttle 2010). This confirms findings by Mitchell and McClure from as early as 1937 (Mitchell & McClure 1937). Today, the I-requirements of food producing animals are given by national scientific bodies such as, Society of Nutrition Physiology (GfE) in Germany (GfE 1995; GfE 1999; GfE 2001; GfE 2006) or the National Research Council (NRC) in the USA (NRC 1994; NRC 1998; NRC 2001; NRC 2005; NRC 2007) and vary between 0.1 and 0.6 mg/kg DM, depending on animal species and category, animal feeding, animal performance/yield, environmental stressors and mainly by the content of glucosinolates in feeds. Under such conditions, depending on the glucosinolate level of feed, the I-requirements may be twice as high (Suttle 2010).

The iodine requirements of animals are comparable with human requirements on a dry matter intake basis- for adults 0.3 - 0.4 mg I/kg DM; for animals 0.1 – 0.6 mg I/kg DM (see Table 2.2.1). The maximum tolerable concentration in feed depends on animal species and varied between 100 and 400 mg/kg feed. The legally permitted iodine content in feed varies between 2 (dairy cows) and 10 mg/kg feed.

4. Iodine Content in Water and Food of Plant Origin

4.1. Drinking Water

Normally, water is not a significant source of minerals. In many countries, the iodine content of drinking water is negligibly low, between 0.2 and 15 µg/L, depending on the location's proximity to the coast (EFSA 2014; Fisher & Delange 1998; Fordyce 2003; McDowell 2003; Röttger et al. 2012; Suttle 2010) and the geological origin of the habitat (Anke, Groppel & Bauch 1993; Anke, Groppel & Scholz 1993). However, in specific locations, water may contain appreciable amounts of specific elements including iodine. Under such conditions, water should be sampled and analyzed routinely to determine whether it is a substantial source of iodine or not.

Near the Alps in the southern part of Germany, Groppel & Anke (1986) measured 1.1 µg iodine/L but, near the sea in North Germany, about 8.0 µg/L in drinking water was measured. Similar data are also given by Felgentraeger (1984) who suggested 11.6 µg iodine/L in the North and 1.6-3.1 µg/L in the South of Germany. Small differences in the iodine content of German beers are also described by Hampel, Kairies & Below (2009)-

3.6 µg/L in the north and 0.5 µg/L in the South. Fordyce (2003) gives 6.4 µg iodine per L water as an average for the British Islands. The range is given between 0.5 and 11.9 µg/L (Fuge 1989). In Finland, the iodine content in tap water varies from 0.3 to 9.1 µg/L (Hasanen 1970). Compared with these data, larger differences are reported between various tap water sources in Denmark (see Table 2.4.1).

From a global point of view, the British Geological Survey (2000) gives a large range of between 0.01 and 70 µg iodine per litre groundwater for drinking with extremes up to 400 µg I/L in saline regions. Large variations in iodine content of water depending on the geographical region are described in the literature on the subject (see Table 2.4.1).

Because of the very low I-concentrations the iodine supply via drinking water was not considered by the authors of dose-response feeding studies with food producing animals (see review by Flachowsky et al. 2014).

Pedersen et al. (1999) analyzed the tap water obtained from 55 different locations in Denmark and found a variation between <1.0 and 139 µg iodine/L. The highest values were found in the tap water of various islands. In Denmark, the high variation of iodine concentrations in tap water is considered as a major determinant of regional differences in iodine intake. The authors found a statistically significant correlation between tap water iodine content today and the urinary iodine excretion measured in 41 towns ($r = 0.68$; $P < 0.001$).

Shen at al. (2011) analyzed 28, 857 water samples of 1978 towns in China. They randomly selected children of 8-10 years and examined the presence of goiter. Of the 1978 towns studied, 488 had iodine levels between 150 and 300 µg/L in drinking water, while in 246 towns, the iodine content was >300 µg/L (compare with data in Table 2.3.1. Of the 56, 751 children examined, goiter prevalence was 6.3% in the areas with drinking water iodine levels of 150 – 300 µg/L and 11.0% in the areas with drinking-water iodine >300 µg/L. In this study, the goiter prevalence increased with water iodine levels.

Table 2.4.1. Water iodine levels by various authors

Region	Mean (µg/L)	Range (µg I/L)	Authors
Samples from UK		1 - 4	Broadhead, Pearson & Wilson (1965)
Goitrous area, Egyptian oases		7 - 18	Coble et al. (1968)
Non goitrous area, Egyptian oases		44 - 100	Coble et al. (1968)
Goitrous area, Sri Lanka		2.2 – 10.1	Mahadeva & Shanmuganathan (1967)
Non-goitrous area, Sri Lanka		19.4 - 183	Mahadeva & Shanmuganathan (1967)
Goitrous areas, India		3 - 16	Ranganathan (1995)
Non-goitrous areas, India		5 – 64	
USA, drinking water	4	< 18	WHO, ICCIDD & UNICEF (2007)
Denmark (55 locations)		<1 - 139	Pedersen et al. (1999)
Sub-Himalayan region (India)	1.5	±0.48	Sharma et al. (1999)
Turkey samples	44	1.8 – 100.4	Unak et al. (1999)
Global, groundwater		0.01 - 70	British Geological Survey (2000)
Sea water	58		British Geological Survey (2000)
Sri Lanka	<84		Fordyce et al. (2000)
Denmark (41 locations)	12.2	2.1 – 30.2	Rasmussen, Larsen & Ovesen (2000)
District Puducherry, India (86 samples)	92	20 - 150	Basu et al. (2007)
Germany, mineral waters	0.5		Hampel, Kairies & Below (2009)
Sea water	>50		Suttle (2010)
Braunschweig, Germany (5 samples)	1.7	1.4 – 2.0	Röttger et al. (2012)

In another Chinese study, Liu et al. (2009) the influence of iodine concentration in drinking water on the intelligence of children in the Tianjin-region. The authors included 1, 229 eight to ten years old school children with a mean intelligence quotient (IQ) of 105.8 (95%: 104.2 – 107.3). The water analyses indicated iodine concentrations in two rural regions as high (137.5) and exceedingly high (234.7 µg/L) in another region. The authors calculated a significant association between the water iodine concentration and the reduction of IQ by an average of about nine points after adjusting for the potential confounding factors. Boiling or heating did not significantly influence the iodine content of water (Rasmussen, Larsen & Ovesen 2000). Furthermore, it seems reasonable to use the same values for iodine in tea as in drinking water. Such data demonstrate that in some regions of the world the I-content of drinking water should not be neglected. Water iodine-concentrations of about 20 µg/L may substantially contribute to meeting the I-requirements of humans (25 – 40% of requirements considering a water intake of about 2 L/day).

Some authors agree that high iodine intake and large thyroid volume may be triggering a process similar to iodine deficiency due to an autoimmune process of lymphoid infiltration of the thyroid resulting in an inhibition of thyroid hormone release (see for example, Henjum et al. 2010; Li et al. 1987; Liu et al. 2009; Lu et al. 2005; Lv et al. 2012; Zimmermann et al. 2005).

In some regions iodine is recommended and used for water disinfection (Backer & Hollowell (2000). The use of iodine for this purpose requires a risk assessment weighing iodine's benefits as a disinfectant and its potential effect on changing thyroid physiology. When using iodine for water disinfection, a maximum recommended dietary dose of 2 mg/day and the maximum recommended duration of use- 3 weeks- should be considered. This value is much higher than the UL, but the application period is limited.

Most authors agree that large geographical variations in iodine concentrations were found in drinking water (see Table 2.4.1). These values are important when calculating the iodine intake of humans and animals.

4.2. Iodized Salt

Iodized salt is a table salt mixed with a small amount of various salts of the element iodine (KI, NaI; NaIO$_3$; KIO$_3$). The objective of iodination of salt is to prevent iodine deficiency in people.

The supplementation of salt with various I-sources was first introduced in Switzerland in 1922, and later in 1924 in the USA (Dasgupta, Liu & Dyke 2008). Since 1976, iodized salt has also been available in Germany. The WHO (WHO, ICCIDD & UNICEF 2007) recommended 20 – 40 mg I-supplementation (mainly KI or KIO$_3$) per kg salt. That means, 1 g salt contains 20 – 40 µg I and these amounts roughly correspond to 10 – 20% of the daily I-need of adults (see Table 1). In the case of high I-intake by foods such as large consumption of milk with >100 µg I/L and high sea fish-consumption, salt with no I content or salt supplemented with lower I-amounts (0 – 20 mg I/kg salt) should be used.

Many factors affect iodine stability in salt, such as the salt's moisture content, ambient humidity, light, heat, impurities in salt, alkalinity or acidity and the form of iodine (iodide or iodate) present in the salt (Kelly 1953). Shawel et al. (2010) conducted field studies in Ethiopia and found a decrease by 57% of iodine because of high temperature and moisture, supplemented as KIO$_3$ from the production side to the consumers. Similar results, mainly from tropical countries, are reported by Assey, Peterson & Greiner (2008); Avinash & Prabha Adhikari (2002); Jooste et al. (1999); Joshi et al. (2007) and some other authors. Further losses are possible during improper storing in households. A creation of adequate storage and handling of iodized salt should prevent possible losses of iodine at the retailer and consumer levels. Against the background of the worldwide public health measure recommending a reduced salt intake (Campbell et al. 2012), an increment of the iodine concentration in salt should be considered.

4.3. Food and Feed of Plant Origin

Most forages contain >100 and <500 µg iodine per kg dry matter (see Table 2.4.2). The vegetation stadium of plants may significantly influence the iodine content of plants as shown in Table 2.4.3 for some forages. A later vegetation stadium effects a strong reduction of the Iodine content per kg DM.

Table 2.4.2. Iodine content of forages (e.g., grass, hay, straw, silages etc.) by various authors

Feed	Origin	Mean (µg/kg DM)	Range (µg/kg DM)	Authors
Hay and straw	Chile		100 – 200	Chilean Iodine Educational Bureau (CIEB) (1952)
Grass hay	Germany	136	SD: 61	Groppel & Anke (1986)
Cereal straw	Germany	368	SD: 283	Groppel & Anke (1986)
Alfalfa, bud	Germany	240		Jeroch, Flachowsky & Weissbach (1993)
Alfalfa, flowering, hay	Germany	240		Jeroch, Flachowsky & Weissbach (1993)
Red clover, bud	Germany	360		Jeroch, Flachowsky & Weissbach (1993)
Red clover, flowering, hay	Germany	290		Jeroch, Flachowsky & Weissbach (1993)
White clover, bud	Germany	430		Jeroch, Flachowsky & Weissbach (1993)
Straw	Germany		320 – 555	Jeroch, Flachowsky & Weissbach (1993)
Grass, fresh	Germany		300 - 470	Jeroch, Flachowsky & Weissbach (1993)
Grass, silage	Germany	173		Schöne, Sporl & Leiterer (2017)

Table 2.4.3. Influence of plant species and vegetation stadium of iodine concentration (µg/kg DM) of various feeds in Germany (Groppel & Anke 1986)

Cut date	24 April	16 June
Meadow fescue	184 ± 24	20 ± 4
Green rye	305 ± 108	43 ± 20
Green wheat	215 ± 53	18 ± 4
Red clover	294 ± 34	104 ± 37
Alfalfa	358 ± 81	149 ± 37

Table 2.4.4. Iodine content of cereals and cereal co-products by various authors

Food/feed	Origin	No. of samples	Mean (µg/kg DM)	Range (µg/kg DM)	Authors
Cereal grains	Chile		40 - 90		Chilean Iodine Educational Bureau (CIEB) (1952)
Barley	Germany		280		Jeroch, Flachowsky & Weissbach (1993)
Barley	Germany	69	95	SD: 69	Groppel & Anke (1986)
Barley	Germany		74	20 - 84	Souci, Fachmann & Kraut (2008)
Maize	Germany	9	44	SD: 18	Groppel & Anke (1986)
Maize	Germany		380		Jeroch, Flachowsky & Weissbach (1993)
Maize	Germany		26	25 - 27	Souci, Fachmann & Kraut (2008)
Millet	Germany		25		Souci, Fachmann & Kraut (2008)
Oats	Germany		80		Souci, Fachmann & Kraut (2008)
Potatoes	Switzerland		16		Haldimann et al. (2005)
Rolled oats	Germany		45	35 - 64	Souci, Fachmann & Kraut (2008)

Table 2.4.4. (Continued)

Food/feed	Origin	No. of samples	Mean (µg/kg DM)	Range (µg/kg DM)	Authors
Rice	Sri Lanka		<58		Fordyce et al. (2000)
Rice	Switzerland	11	333	11 – 934	Haldimann et al. (2005)
Rice	Germany		22		Souci, Fachmann & Kraut (2008)
Rice, polished	Germany		19		Souci, Fachmann & Kraut (2008)
Rye	Germany		200		Jeroch, Flachowsky & Weissbach (1993)
Rye	Germany		72		Souci, Fachmann & Kraut (2008)
Winter wheat	Germany		360		Jeroch, Flachowsky & Weissbach (1993)
Wheat	Switzerland	11	35	11 - 47	Haldimann et al. (2005)
Wheat	Germany		67	6 - 87	Souci, Fachmann & Kraut (2008)
Wheat bran	Germany		320		Jeroch, Flachowsky & Weissbach (1993)
Wheat bran	Germany		310		Souci, Fachmann & Kraut (2008)

Most cereal seeds and by-products of cereals contain less iodine than forages. The large ranges within the data of some cereals or co-products from cereals between various authors cannot be explained in detail. The iodine content of soil and water may possibly influence the iodine-concentration of cereals and co-products, too.

Less data are available for legumes and legume co-products (see Table 2.4.5). The values are comparable with cereals (compare Tables 2.4.4 and 2.4.5). Soybeans seem to be the richest source of iodine.

Low values are also given for fruits and vegetables (Table 2.4.6). On the other hand, mushrooms and nuts are richer in iodine (Table 2.4.6).

Table 2.4.5. Iodine content of legumes and their by-products by various authors

Food/feed	Origin	Mean (µg/kg DM)	Range (µg/kg DM)	Authors
Bean	Germany	18		Souci, Fachmann & Kraut (2008)
Peas, green	Germany	42	23 – 60	Souci, Fachmann & Kraut (2008)
Peas, seed	Germany	22		Souci, Fachmann & Kraut (2008)
Peas	Germany	150		Jeroch, Flachowsky & Weissbach (1993)
Soybean, dry	Germany	63		Souci, Fachmann & Kraut (2008)
Oilseed meals	Chile	100 - 200		Chilean Iodine Educational Bureau (CIEB) (1952)
Rapeseed (extr.)	Germany	67	SD: 30	Groppel & Anke (1986)
Soybean	Germany	500		Jeroch, Flachowsky & Weissbach (1993)
Soybean meal (extr.)	Germany	97	SD. 33	Groppel & Anke (1986)

The iodine content of potatoes is given with 160 – 220 µg/kg DM (Jeroch, Flachowsky & Weissbach 1993). Haldimann et al. (2005) state for potatoes an average of 16 µg I/kg and a range between 4 and 26 µg/kg dry weight samples.

4.4. Seaweeds

In Asia, seaweeds have a long history and are a constant factor of nutrition and therefore, an important iodine source (Packer, Harris & L. 2016). On average, the Japanese eat 1.4 kg seaweed per person per year (Burtin 2003). Dietary seaweeds include species that are directly used for human consumption, such as *Codium fragile, Dictyopteris divaricata, Ulva pertusa, Ulva rigida, Monostroma sp. Entoromorpha sp., Laminaria sp.* among others (Burtin 2003; Gupta & Abu-Ghannam 2011; Hou et al. 1997; Kerkvliet 2001; MacArtain et al. 2007; Teas et al. 2004; van der Spiegel, Noordam & van der Fels-Klerx 2013).

Table 2.4.6. Iodine content of fruits and vegetables

Food	Origin	Number of samples	Mean (µg/kg DM)	Range (µg/kg DM)	Authors
Fruits	Chile		18	10 - 29	Chilean Iodine Educational Bureau (CIEB) (1952)
Fruits	UK			<20 - 80	Wenlock et al. (1982)
Fruits	USA		<30		Pennington et al. (1995)
Fruits	Switzerland, 2000-2001		3	0.3 - 13	Haldimann et al. (2005)
Fresh fruits	Switzerland	62	18	2 – 75	Haldimann et al. (2005)
Vegetables	Chile		29	12 - 201	Chilean Iodine Educational Bureau (CIEB) (1952)
Vegetables	Finnland		<10		Varo et al. (1982)
Vegetables	UK			<20 - 280	Wenlock et al. (1982)
Vegetables	USA		<10		Pennington et al. (1995)
Vegetables	Switzerland; 2000 – 2001		5	1 – 22	Haldimann et al. (2005)
Fresh vegetables	Switzerland	36	47	9 - 203	Haldimann et al. (2005)
Mushrooms	Switzerland	10	211	44 - 426	Haldimann et al. (2005)
Nuts	Switzerland	13	218	20 - 374	Haldimann et al. (2005)

Seaweeds are rich in minerals, in some cases, up to 36% of dry matter (Burtin 2003). Laminaria is considered as the iodine richest seaweed with a value of 1, 500 – 8, 000 mg I/kg DM (Burtin 2003; Kerkvliet 2001). Teas et al. (2004) analyzed 12 different species of seaweeds and found a range of Iodine from 16 mg (*Porphyra tenera*) up to 8 165 mg/kg (*Laminaria digitata*). Some authors (e.g., Hetzel & Maberly 1986) and international organizations (e.g., National Academy of Sciences (NAS) 1980) give the iodine content of seaweed with one to two grams per kg substance. Hou et al. (1997) determined 734 mg I/kg wet *Laminaria japonica*; all other analyzed algae contained less iodine.

On the other hand, seaweeds can also accumulate heavy metals (van der Spiegel, Noordam & van der Fels-Klerx 2013). The uptake of heavy metals for *Palmaria palmata* was found to decrease in the order Pb > Cd > Cu > Ni (Prasher et al. 2004).

Compared with seaweed, algae and other fresh water plants, also called duckweed species, are rich in protein and amino acids (Becker 2007; Kovač et al. 2013; Packer, Harris & L. 2016; Priyadarshani & Rath 2012) and may contain some heavy metals, but they are poor in iodine and, therefore, cannot be considered as reliable or significant iodine sources (van der Spiegel, Noordam & van der Fels-Klerx 2013).

CONCLUSION

The world has achieved considerable progress in the global human iodine nutrition and the Iodine status worldwide, although about 30 countries are still considered as iodine-deficient (<99 µg I/L urine). This progress has been achieved through programs of salt iodination, but also by iodine supplementation of food and animal feed. Many years of research have enhanced our knowledge about iodine dynamics and implications for human health but has also shown that our knowledge is incomplete, and raised questions that still need to addressed.

Iodine is an essential trace element for human nutrition and consequently human health. I-intake has been correlated with vital processes in the body negatively. For example, clinical studies show that iodine deficiency has deleterious effects on embryo development, brain development, and thyroid health. Severe lack of iodine during pregnancy has also been found to affect mental development.

Human I-requirements vary between 40 and >200 µg/day depending on many influencing factors. There is a small range between human requirements of iodine and the upper level (about 1: 2.5-3). Drinking water, iodized salt, and some food plants may contribute to meeting iodine requirements of humans and animals. For example, various seaweeds are very rich in iodine (up to 8 000 mg/kg DM) and may contribute to satisfying human and animal requirements. This chapter focused on non-meat I-sources. Much of the gains in iodine status globally however, is due to advances related to meat, poultry, and to a lesser extent, fish. These are discussed in Chapter Four of this volume.

REFERENCES

Andersen, NL, Fagt, S, Groth, MV, Hartkopp, HB, Moller, A, Ovesen, L & Wanning, DL 1996, 'Danskernes kostvaner 1995,' *Levnedsmiddelstyrelsen Publ. No. 235*.

Andersson, M, de Benoist, B, Delange, F, Zupan, J & Secretariat, W 2007, 'Prevention and control of iodine deficiency in pregnant and lactating women and in children less than 2-years-old: conclusions and recommendations of the Technical Consultation,' *Public Health Nutrition*, vol. 10, no. 12A, pp. 1606-1611. Available from: <Go to ISI>://WOS:000252677600015.

Andersson, M, de Benoist, B & Rogers, L 2010, 'Epidemiology of iodine deficiency: Salt iodisation and iodine status,' *Best Practice & Research Clinical Endocrinology & Metabolism*, vol. 24, no. 1, pp. 1-11. Available from: <Go to ISI>://WOS:000275917500002.

Andersson, M, Karumbunathan, V & Zimmermann, MB 2012, 'Global iodine status in 2011 and trends over the past decade,' *Journal of*

Nutrition, vol. 142, no. 4, pp. 744-750. Available from: http://jn.nutrition.org/content/142/4/744.abstract.

Anke, M, Groppel, B & Bauch, K-H 1993, 'Iodine in the Food Chain,' in F Delange, JT Dunn & D Glinoer, (eds), *Iodine Deficiency in Europe: A Continuing Concern*, pp. 151-158. Springer US, Boston, MA.

Anke, M, Groppel, B & Scholz, E 1993, *Iodine in the food chain*, Boston, MA.

Assey, VD, Peterson, S & Greiner, T 2008, 'Sustainable universal salt iodization in low-income countries - time to re-think strategies?,' *European Journal of Clinical Nutrition*, vol. 62, no. 2, pp. 292-294. Available from: <Go to ISI>://WOS:000252932900019.

Australian Government, *National Health and Medical Research Council (NHMCR)Mo 2006*, 'Nutrient Reference Values for Australia and New Zealand Including Recommended Dietary Intakes (last update 2017).' Available from: https://www.nhmrc.gov.au/guidelines-publications/n35-n36-n37.

Avinash, KR & Prabha Adhikari, MR 2002,''Iodine content of various salt samples sold in Mangalore--a coastal city with endemic goitre,' *J Assoc Physicians India*, vol. 50, pp. 1146-8. Available from: http://www.ncbi.nlm.nih.gov/pubmed/12516697.

Backer, H & Hollowell, J 2000, 'Use of iodine for water disinfection: Iodine toxicity and maximum recommended dose,' *Environmental Health Perspectives*, vol. 108, no. 8, pp. 679-684. Available from: <Go to ISI>://WOS:000089296800015.

Basu, S, Mohanty, B, Sonali Sarkar, S & Kumar, GS 2007, 'Estimation of iodine levels in drinking water in Puducherry district,' *Biomedical Research 2007*, vol. 18, no. 3, pp. 171-173.

Becker, EW 2007, 'Micro-algae as a source of protein,' *Biotechnology Advances*, vol. 25, no. 2, pp. 207-210. Available from: <Go to ISI>://WOS:000244412000008.

British Geological Survey 2000, '*Water Quality Fact Sheet: Iodine,*' pp. 1-4.

Broadhead, GD, Pearson, IB & Wilson, GM 1965, 'Seasonal Changes in Iodine Metabolism. I. Iodine Content of Cows Milk,' *Bmj-British*

Medical Journal, vol. 1, no. 5431, pp. 343-+. Available from: <Go to ISI>://WOS:A19656118700009.

Burtin, P 2003, 'Nutritional Value of Seaweeds,' *Electronic Journal of Environmental, Agricultural and Food Chemistry*, vol. 2, no. 4, pp. 498-503.

Campbell, N, Dary, O, Cappuccio, FP, Neufeld, LM, Harding, KB & Zimmermanne, MB 2012, 'Collaboration to optimize dietary intakes of salt and iodine: a critical but overlooked public health issue,' *Bulletin of the World Health Organization*, vol. 90, no. 1, pp. 73-74. Available from: <Go to ISI>://WOS:000299913600028.

Chacón Villanueva, C & Agency, CFS 2016, *Total diet study of iodine and the contribution of milk in the exposure of the catalan population, 2015*, MoH Government of Catalonia, Public Health Agency of Catalonia.

Chilean Iodine Educational Bureau 1952, *Iodine content of foods; annotated bibliography, 1825-1951, with review and tables*, London. Available from: /z-wcorg/.

Coble, Y, Davis, J, Schulert, A, Heta, F & Awad, AY 1968, 'Goiter and iodine deficieny in gyptian oases,' *Am J Clin Nutr*, vol. 21, no. 4, pp. 277-83. Available from: http://www.ncbi.nlm.nih.gov/pubmed/4171731.

DACH 2000, *Referenzwerte für die Nährstoffzufuhr*, [Reference values for nutrient intake] Deutsche Gesellschaft für Ernährung, Österreichische Gesellschaft für Ernährung, Schweizerische Gesellschaft für Ernährungsforschung, Schweizerische Vereinigung für Ernährung, Umschau/Braus Verlag, Frankfurt am Main.

Dahl, L, Opsahl, JA, Meltzer, HM & Julshamn, K 2003, ‚Iodine concentration in Norwegian milk and dairy products,' *British Journal of Nutrition*, vol. 90, no. 03, pp. 679-685. Available from: http://dx.doi.org/10.1017/S0007114503001740. [2003].

Dasgupta, PK, Liu, Y & Dyke, JV 2008, 'Iodine nutrition: iodine content of iodized salt in the United States,' *Environ Sci Technol*, vol. 42, no. 4, pp. 1315-23. Available from: http://www.ncbi.nlm.nih.gov/pubmed/18351111.

de Benoist, B, McLean, E, Andersson, M & Rogers, L 2008, 'Iodine deficiency in 2007: Global progress since 2003,' *Food and Nutrition Bulletin*, vol. 29, no. 3, pp. 195-202. Available from: <Go to ISI>://WOS:000259946500005.

Decuypere, E, Van As, P, Van der Geyten, S & Darras, VM 2005, 'Thyroid hormone availability and activity in avian species: A review,' *Domestic Animal Endocrinology*, vol. 29, no. 1, pp. 63-77. Available from: http://www.sciencedirect.com/science/article/pii/S073972400 5000445.

EFSA 2005, 'Opinion of the Scientific Panel on Additives and Products or Substances used in Animal Feed on the request from the Commission on the use of iodine in feedingstuffs,' *EFSA Journal*, vol. 168, no. 2, pp. 1-42.

EFSA 2013a, 'Scientific Opinion on the safety and efficacy of iodine compounds (E2) as feed additives for all animal species: calcium iodate anhydrous, based on a dossier submitted by Calibre Europe SPRL/BVBA,' *EFSA Journal*, vol. 11, no. 2, p. 3100 [34pp.].

EFSA 2013b, 'Scientific Opinion on the safety and efficacy of Iodine compounds (E2) as feed additives for all species: calcium iodate anhydrous (coated granulated preparation), based on a dossier submitted by Doxal Italia S.p.A.,' *EFSA Journal* vol. 11(3), no. 3178, p. 36 pp.

EFSA 2014, 'Scientific Opinion on Dietary Reference Values for iodine,' *EFSA Journal*, vol. 12(5), no. 3660, p. 57 pp.

Felgentraeger, HJ 1984, 'Zum Jodgehalt der Umwelt in der DDR und seine Beziehungen zum Gesundheitszustand der Bevölkerung,' *Zeitschrift der Gesellschaft für Hygiene* no. 30, pp. 154-155.

Fisher, DA & Delange, FM 1998, 'Thyroid hormone and iodine requirements in man during brain development,' in JB Stannury, FM Delange, JT Dunn & CS Pandav, (eds), *Iodine in Pregnancy*. Oxford Univ. Publ., New Delhi.

Flachowsky, G, Franke, K, Meyer, U, Leiterer, M & Schöne, F 2014, 'Influencing factors on iodine content of cow milk,' *European Journal of Nutrition*, vol. 53, no. 2, pp. 351-365. Available from: <Go to ISI>://WOS:000331710600001.

Fordyce, FM 2003, 'Database of the Iodine Content of Food and Diets Populated with Data from Published Literature.,' *British Geological Survey Commissioned Report, CR/03/84N.*

Fordyce, FM, Johnson, CC, Navaratna, UR, Appleton, JD & Dissanayake, CB 2000, 'Selenium and iodine in soil, rice and drinking water in relation to endemic goitre in Sri Lanka,' *Sci Total Environ*, vol. 263, no. 1-3, pp. 127-41.

Fuge, R 1989, 'Iodine in waters: possible links with endemic goitre,' *Applied Geochemistry*, vol. 4, pp. 203-208.

GfE 1995, *Empfehlungen zur Energie- und Nährstoffversorgung der Mastrinder* [Recommendations for the energy and nutrient supply of the beef cattle] DLG-Verlag (1995).

GfE 1999, *Empfehlungen zur Energie- und Nährstoffversorgung der Legehennen und Masthühner* [Recommendations for the energy and nutrient supply of laying hens and chickens for fattening] *(Broiler) 1999.*

GfE 2001, *Empfehlungen zur Energie und Nährstoffversorgung der Milchkühe und Aufzuchtrinder*, [Recommendations on the energy and nutrient supply of dairy cows and breeding cattle] DLG-Verlag, Frankfurt am Main.

GfE 2006, *Empfehlungen zur Energie- und Nährstoffversorgung bei Schweinen.* [Recommendations for energy and nutrient supply in pigs.]

Groppel, B & Anke, M 1986, 'Iodine supply of farm animals in the German Democratic Republic,' *Tierzucht (German D. R.)*, vol. 40, no.5, pp. 236-238.

Gunnarsdottir, I & Dahl, L 2012, 'Iodine intake in human nutrition: a systematic literature review,' *Food Nutr Res*, vol. 56. Available from: http://www.ncbi.nlm.nih.gov/pubmed/23060737.

Gupta, S & Abu-Ghannam, N 2011, 'Recent developments in the application of seaweeds or seaweed extracts as a means for enhancing the safety and quality attributes of foods,' *Innovative Food Science & Emerging Technologies*, vol. 12, no. 4, pp. 600-609. Available from: <Go to ISI>://WOS:000297232000026.

Haldimann, M, Alt, A, Blanc, A & Blondeau, K 2005, 'Iodine content of food groups,' *Journal of Food Composition and Analysis*, vol. 18, no. 6, pp. 461-471. Available from: <Go to ISI>://WOS:000227850800001.

Hampel, R, Kairies, J & Below, H 2009, 'Beverage iodine levels in Germany,' *European Food Research and Technology*, vol. 229, no. 4, pp. 705-708. Available from: <Go to ISI>://WOS:000268275900019.

Hasanen, E 1970, 'Iodine Content of Drinking Water and Diseases of Circulatory System,' *Annales Medicinae Experimentalis Et Biologiae Fenniae*, vol. 48, no. 2, pp. 117-&. Available from: <Go to ISI>://WOS:A1970G801100009.

Henjum, S, Barikmo, I, Gjerlaug, AK, Mohamed-Lehabib, A, Oshaug, A, Strand, TA & Torheim, LE 2010, 'Endemic goitre and excessive iodine in urine and drinking water among Saharawi refugee children,' *Public Health Nutrition*, vol. 13, no. 9, pp. 1472-1477. Available from: <Go to ISI>://WOS:000282026600023.

Hester, P 2017, *Egg Innovations and Strategies for Improvements*, 1st ed.

Hetzel, BS & Maberly, GF 1986, 'Iodine,' in W Mertz, (ed) *Trace elements in human and animal nutrition*, 5 edn, pp. 139-208. Academic Press New York, NY.

Hou, XL, Chai, CF, Qian, QF, Yan, XJ & Fan, X 1997, 'Determination of chemical species of iodine in some seaweeds .1.,' *Science of the Total Environment*, vol. 204, no. 3, pp. 215-221. Available from: <Go to ISI>://WOS:A1997XY61600002.

Institute of Medicine & (US) Panel on Micronutrients 2001. *'Dietary Reference Intakes for Vitamin A, Vitamin K, Arsenic, Boron, Chromium, Copper, Iodine, Iron, Manganese, Molybdenum, Nickel, Silicon, Vanadium, and Zinc.'* (US) National Academies Press Washington (DC). Available from: https://www.ncbi.nlm.nih.gov/books/NBK222310/ doi: 10.17226/10026.

Joint FAO/WHO-Expert Committee on Food Additives 1990. 'Toxicological evaluation of certain food additives and contaminants,' *WHO Food Additives Series*, vol. 26.

Jeroch, H, Flachowsky, G & Weissbach, F 1993, *Futtermittelkunde*, [Feed Science,] Elsevier, München.

Johner, SA, von Nida, K, Jahreis, G & Remer, T 2012, 'Time trends and seasonal variation of iodine content in German cow's milk - investigations from Northrhine-Westfalia,' *Berliner und Munchener Tierarztliche Wochenschrift*, vol. 125, no. 1-2, pp. 76-82. Available from: <Go to ISI>://WOS:000299581400012.

Jooste, PL, Weight, MJ, Locatelli-Rossi, L & Lombard, CJ 1999, 'Impact after 1 year of compulsory iodisation on the iodine content of table salt at retailer level in South Africa,' *Int J Food Sci Nutr*, vol. 50, no. 1, pp. 7-12. Available from: http://www.ncbi.nlm.nih.gov/pubmed/10435116.

Joshi, AB, Banjara, MR, Bhatta, LR, Rikimaru, T & Jimba, M 2007, 'Insufficient level of iodine content in household powder salt in Nepal,' *Nepal Med Coll J*, vol. 9, no. 2, pp. 75-8. Available from: http://www.ncbi.nlm.nih.gov/pubmed/17899952.

Kelly, FC 1953, 'Studies on the stability of iodine compounds in iodized salt,' *Bull World Health Organ*, vol. 9, no. 2, pp. 217-30. Available from: http://www.ncbi.nlm.nih.gov/pubmed/13094510.

Kerkvliet, JD 2001, 'Algen en zeewieren als levensmiddel : een overzicht,' ['Algae and seaweeds as food: an overview,'] *De ware(n)-chemicus: Nederlands tijdschrift voor algemeen levensmiddelenonderzoek*, vol. 31, no. 2.

Kovač, DJ, Simeunović, JB, Babić, OB, Mišan, AČ & Milovanović, IL 2013, 'Algae in food and feed,' *Food & Feed Research*, vol. 40, no. 1, pp. 21-32.

Küpper, FC, Feiters, MC, Olofsson, B, Kaiho, T, Yanagida, S, Zimmermann, MB, Carpenter, LJ, Luther, GW, Lu, ZL, Jonsson, M & Kloo, L 2011, 'Commemorating Two Centuries of Iodine Research: An Interdisciplinary Overview of Current Research,' *Angewandte Chemie-International Edition*, vol. 50, no. 49, pp. 11598-11620. Available from: <Go to ISI>://WOS:000298084900006.

Li, M, Qu, CG, Qian, QD, Jia, QZ, Eastman, CJ, Collins, JK, Maberly, GF, Liu, DR, Zhang, PY, Zhang, CD, Wang, HX, Boyages, SC & Jupp, JJ 1987, 'Endemic Goiter in Central China Caused by Excessive Iodine

Intake,' *Lancet*, vol. 2, no. 8553, pp. 257-258. Available from: <Go to ISI>://WOS:A1987J397800013.

Liu, HL, Lam, LT, Zeng, Q, Han, SQ, Fu, G & Hou, CC 2009, 'Effects of drinking water with high iodine concentration on the intelligence of children in Tianjin, China,' *Journal of Public Health*, vol. 31, no. 1, pp. 32-38. Available from: <Go to ISI>://WOS:000263835600008.

Lu, YL, Wang, NJ, Zhu, L, Wang, GX, Wu, H, Kuang, L & Zhu, WM 2005, 'Investigation of iodine concentration in salt, water and soil along the coast of Zhejiang, China,' *J Zhejiang Univ Sci B*, vol. 6, no. 12, pp. 1200-5.

Lv, S, Zhao, J, Xu, D, Chong, Z, Jia, L, Du, Y, Ma, J & Rutherford, S 2012, 'An epidemiological survey of children's iodine nutrition and goitre status in regions with mildly excessive iodine in drinking water in Hebei Province, China,' *Public Health Nutrition*, vol. 15, no. 7, pp. 1168-1173. Available from: Cambridge Core. Available from: https://www.cambridge.org/core/article/an-epidemiological-survey-of-childrens-iodine-nutrition-and-goitre-status-in-regions-with-mildly-excessive-iodine-in-drinking-water-in-hebei-province-china/A213E89E38CDC0A62A8F07D0B89E4FF6.

MacArtain, P, Gill, CI, Brooks, M, Campbell, R & Rowland, IR 2007, 'Nutritional value of edible seaweeds,' *Nutr Rev*, vol. 65, no. 12 Pt 1, pp. 535-43.

Mahadeva, K & Shanmuganathan, SS 1967, 'Problem of Goitre in Ceylon,' *British Journal of Nutrition*, vol. 21, no. 2, pp. 341-+. Available from: <Go to ISI>://WOS:A19679518800011.

McDowell, LR 2003, *Minerals in animal and human nutrition*.

Mitchell, JH & McClure, FJ 1937, 'Mineral nutrition of farm animals. National Research Council. Washington, DC,' *Bull. No. 99*.

National Academy of Sciences (NAS) 1980; *'Drinking Water and Health,'* Washington, DC, The National Academies Press.

Nordic Nutrition Recommendations, 2012, *Integrating nutrition and physical activity*, Copenhagen.

NRC 1994, *Nutrient Requirements of Poultry: Ninth Revised Edition*, The National Academies Press.

NRC 1998, *Nutrient Requirements of Swine: 10th Revised Edition*, The National Academies Press.

NRC 2001, *Nutrient Requirements of Dairy Cattle: Seventh Revised Edition, 2001*, The National Academies Press.

NRC 2005, *'Mineral tolerance of animals.'* Available from: https://www.nap.edu/read/11309/chapter/1.

NRC 2007, *Nutrient Requirements of Small Ruminants: Sheep, Goats, Cervids, and New World Camelids*, The National Academies Press.

Opalinski, S 2017, 'Supplemental iodin,' in PY Hester, (ed) *Egg innovations and strategies for improvements*, pp. 393-402.

Packer, MA, Harris, GC & L., AS 2016, 'Food and feed applications of algae,' in F Bux & Y Chisti, (eds), *Algae Biotechnology – Products and Processes Part of the series Green Energy and Technology*, pp. 217 - 247. Springer.

Pearce, EN, Andersson, M & Zimmermann, MB 2013, 'Global Iodine Nutrition: Where Do We Stand in 2013?,' *Thyroid*, vol. 23, no. 5, pp. 523-528. Available from: <Go to ISI>://WOS:000318220700001.

Pedersen, KM, Laurberg, P, Nohr, S, Jorgensen, A & Andersen, S 1999, 'Iodine in drinking water varies by more than 100-fold in Denmark. Importance for iodine content of infant formulas,' *European Journal of Endocrinology*, vol. 140, no. 5, pp. 400-403. Available from: <Go to ISI>://WOS:000080508500006.

Pennington, JAT, Schoen, SA, Salmon, GD, Young, B, Johnson, RD & Marts, RW 1995, 'Composition of core foods of the US food supply, 1982-1991,' *Journal of Food Composition Analysis*, no. 8, pp. 171-217.

Prasher, SO, Beaugeard, M, Hawari, J, Bera, P, Patel, RM & Kim, SH 2004, 'Biosorption of heavy metals by red algae (Palmaria palmata),' *Environmental Technology*, vol. 25, no. 10, pp. 1097-1106. Available from: <Go to ISI>://WOS:000225382900001.

Preedy, VR, Burrow, GN & Watson, R 2009, *Comprehensive handbook of iodine: nutritional, biochemical, pathological and therapeutic aspects*.

Priyadarshani, I & Rath, B 2012, 'Commercial and industrial applications of micro algae – A review,' *Journal of Algal Biomass Utilization* vol. 3, no. 4, pp. 89-100.

Ranganathan, S 1995, 'Iodised salt is safe,' *Indian J Public Health*, vol. 39, no. 4, pp. 164-71. Available from: http://www.ncbi.nlm.nih.gov/pubmed/8690505.

Rasmussen, LB, Larsen, EH & Ovesen, L 2000, 'Iodine content in drinking water and other beverages in Denmark,' *European Journal of Clinical Nutrition*, vol. 54, no. 1, pp. 57-60. Available from: <Go to ISI>://WOS:000085395400011.

Röttger, AS 2012, *The effect of various iodine sources and levels on the performance and the iodine transfer in poultry products and tissues*, thesis, Tierärztliche Hochschule Hannover, Germany.

Röttger, AS, Halle, I, Wagner, H, Breves, G, Daenicke, S & Flachowsky, G 2012, 'The effects of iodine level and source on iodine carry-over in eggs and body tissues of laying hens,' *Archives of Animal Nutrition*, vol. 66, no. 5, pp. 385-401. Available from: <Go to ISI>://WOS:000308726000004.

Sager, M 2011, *Major and trace elements in hens' eggs from Austria*.

SCF 2002, *Opinion of the Scientific Committee on Food on the Tolerable Upper Intake Level of Iodine* in http://ec.europa.eu/food/fs/sc/scf/out146_en.pdf.

Schöne, F & Rajendram, R 2009, 'Iodine in farm animals,' in VR Preedy, GN Burrow & RR Watson, (eds), *Comprehensive Handbook of Iodine. Nutritional, Pathological and Therapeutic Aspects*, pp. 151-70 Oxford.

Schöne, F, Sporl, K & Leiterer, M 2017, 'Iodine in the feed of cows and in the milk with a view to the consumer's iodine supply,' *Journal of Trace Elements in Medicine and Biology*, vol. 39, pp. 202-209. Available from: <Go to ISI>://WOS:000390505900029.

Sharma, SK, Chelleng, PK, Gogoi, S & Mahanta, J 1999, 'Iodine status of food and drinking water of a sub-Himalayan zone of India,' *International Journal of Food Sciences and Nutrition*, vol. 50, no. 2, pp. 95-98. Available from: <Go to ISI>://WOS:000079498200002.

Shawel, D, Hagos, S, Lachat, CK, Kimanya, ME & Kolsteren, P 2010, 'Post-production Losses in Iodine Concentration of Salt Hamper the Control of Iodine Deficiency Disorders: A Case Study in Northern Ethiopia,'

Journal of Health Population and Nutrition, vol. 28, no. 3, pp. 238-244. Available from: <Go to ISI>://WOS:000279758900005.

Shen, HM, Liu, SJ, Sun, DJ, Zhang, SB, Su, XH, Shen, YF & Han, HP 2011, 'Geographical distribution of drinking-water with high iodine level and association between high iodine level in drinking-water and goitre: a Chinese national investigation,' *British Journal of Nutrition*, vol. 106, no. 2, pp. 243-247. Available from: <Go to ISI>://WOS: 000291986200011.

Slupczynska, M, Jamroz, D, Orda, J & Wiliczkiewicz, A 2014, 'Effect of various sources and levels of iodine, as well as the kind of diet, on the performance of young laying hens, iodine accumulation in eggs, egg characteristics, and morphotic and biochemical indices in blood,' *Poultry Science*, vol. 93, no. 10, pp. 2536-2547. Available from: <Go to ISI>://WOS:000343697800013.

Souci, SW, Fachmann, W & Kraut, H 2008, [*Food Composition and Nutrition Tables*]: *Die Zusammensetzung der Lebensmittel, Nährwert-Tabellen La composition des aliments Tableaux des valeurs nutritives*, 7 edn, Wissenschaftliche Verlagsgesellschaft.

Suttle, N 2010, 'Mineral nutrition of livestock,' *Mineral nutrition of livestock*, no. Ed.4, p. viii + 587. Available from: <Go to ISI>://CABI: 20103291114.

Teas, J, Pino, S, Critchley, A & Braverman, LE 2004, 'Variability of iodine content in common commercially available edible seaweeds,' *Thyroid*, vol. 14, no. 10, pp. 836-841. Available from: <Go to ISI>://WOS:000225071800008.

Unak, P, Darcan, S, Yurt, F, Biber, Z & Coker, M 1999, 'Determination of iodide amounts in urine and water by isotope dilution analysis,' *Biological Trace Element Research*, vol. 71-2, pp. 463-470. Available from: <Go to ISI>://WOS:000084610100051.

van der Haar, F, Gerasimov, G, Tyler, VQ & Timmer, A 2011, 'Universal salt iodization in the Central and Eastern Europe, Commonwealth of Independent States (CEE/CIS) Region during the decade 2000-09: experiences, achievements, and lessons learned,' *Food Nutr Bull*, vol.

32, no. 4 Suppl, pp. S175-294. Available from: http://www.ncbi.nlm.nih.gov/pubmed/22416358.

van der Spiegel, M, Noordam, MY & van der Fels-Klerx, HJ 2013, 'Safety of Novel Protein Sources (Insects, Microalgae, Seaweed, Duckweed, and Rapeseed) and Legislative Aspects for Their Application in Food and Feed Production,' *Comprehensive Reviews in Food Science and Food Safety*, vol. 12, no. 6, pp. 662-678. Available from: <Go to ISI>://WOS:000325631600005.

Varo, P, Saari, E, Paaso, A & Koivistoinen, P 1982, 'Iodine in Finnish Foods,' *International Journal for Vitamin and Nutrition Research*, vol. 52, no. 1, pp. 80-89. Available from: <Go to ISI>://WOS:A1982NM92100013.

Wenlock, RW, Buss, DH, Moxon, RE & Bunton, NG 1982, 'Trace nutrients. 4. Iodine in British food,' *Br J Nutr*, vol. 47, no. 3, pp. 381-90. Available from: http://www.ncbi.nlm.nih.gov/pubmed/7082612.

WHO 2004, 'Iodine status worldwide: WHO Global Database on Iodine Deficiency.'

WHO, ICCIDD & UNICEF 2001, *Assessment of iodine deficiency disorders and monitoring their elimination*, p. 107.

WHO, ICCIDD & UNICEF 2007, *Assessment of iodine deficiency disorders and monitoring their elimination. A guide for programme managers. 3rd. ed.*

Zava, TT & Zava, DT 2011, 'Assessment of Japanese iodine intake based on seaweed consumption in Japan: A literature-based analysis,' *Thyroid Res*, vol. 4, p. 14. Available from: http://www.ncbi.nlm.nih.gov/pubmed/21975053.

Zimmermann, MB 2009, 'Iodine Deficiency,' *Endocrine Reviews*, vol. 30, no. 4, pp. 376-408. Available from: <Go to ISI>://WOS:000266731700003.

Zimmermann, MB & Andersson, M 2011, 'Prevalence of iodine deficiency in Europe in 2010,' *Annales D Endocrinologie*, vol. 72, no. 2, pp. 164-166. Available from: <Go to ISI>://WOS:000290841000024.

Zimmermann, MB & Andersson, M 2012, 'Assessment of iodine nutrition in populations: past, present, and future,' *Nutrition Reviews*, vol. 70, no. 10, pp. 553-570. Available from: <Go to ISI>://WOS: 000309448300001.

Zimmermann, MB, Ito, Y, Hess, SY, Fujieda, K & Molinari, L 2005, 'High thyroid volume in children with excess dietary iodine intakes.,' *American Journal of Clinical Nutrition*, vol. 81, no. 4, pp. 840-844. Available from: <Go to ISI>://WOS:000230474400033.

In: Agriculture, Food, and Food Security
Editor: Clinton Lloyd Beckford

ISBN: 978-1-53613-483-4
© 2018 Nova Science Publishers, Inc.

Chapter 3

SIMULATING THE FUTURE OF FOOD DESERTS IN YPSILANTI, MICHIGAN USING MARKOV CHAINS AND CELLULAR AUTOMATA

Hugh Semple[*]
Department of Geography and Geology
Eastern Michigan University, Ypsilanti, Michigan, US

1. INTRODUCTION

Over the last two decades, the topic of food deserts has been extensively discussed in the food insecurity literature (Walker, Keane & Burke 2010; Beaulac, Kristjansson & Cummins 2009). Although they have been variously defined, food deserts are generally thought of as low-income areas in urban and rural areas where there is a lack of access to fresh produce, milk, meat, and other healthy groceries within a convenient distance to

---

[*] Corresponding author email: hsemple@emich.edu

people's homes (Dutko, Ver Ploeg & Farrigan 2012; Smith & Morton 2009). Also see Beckford and Igbokwe in Chapter 6 of this volume for a more detailed discussion of food deserts.

In the context of urban food deserts, many people do not have ready access to private transportation, so getting to food stores some distance from their homes can be difficult. This is particularly the case, if they do not live along bus routes. Consequently, people rely on nearby convenience stores for much of their food needs. However, convenience stores typically do not carry fresh produce, milk, meat, and other healthy groceries in large quantities. As a result, the metaphor "food desert" has been used to describe these low-income areas where there is a preponderance of convenience stores selling non-nutritious food, and a paucity of supermarkets selling nutritious foods.

In addition to the preponderance of convenience stores in food desert areas, studies have also identified a tendency for these areas to have higher exposure to fast food restaurants and liquor stores compared to other locations in a city (Block, Scribner & DeSalvo 2004; Fraser et al. 2010; Alameda County Public Health Department 2008, pp 97-101).

As a result of the high dependence of people in food deserts on convenience stores and fast food restaurants for much of their food supplies, research has also shown that people living in these areas tend to have higher rates of childhood and adult obesity and diabetes compared to surrounding areas (Schafft, Jensen & Hinrichs 2009; Babey et al. 2008). Despite academic debate on the connection between food desert, obesity and diabetes, many health departments and officials are of the view that reducing the spread of food deserts can play a significant role in reducing obesity and diabetes rates in inner city areas, as well as improving the overall health of these communities (Babey et al. 2008; Alameda County Public Health Department 2008, pp 97-101).

Urban food deserts are mostly found in inner city areas that have experienced outmigration and economic decline as a result of suburbanization in the post-World War II period (Larsen & Gilliland 2008). They are also found in older industrial areas that have experienced manufacturing decline due to globalization (McClintock 2008). In both situations, as capital moved to newer locations where returns on investment were higher, people fled formerly attractive areas, which, over time, became characterized by economic decline and urban blight. Food deserts are another consequence of capital flight because supermarkets that once serviced these areas with fresh produce and meat have moved on to more profitable locations (Semple & Giguere 2017).

The vast majority of studies on food deserts in the US have investigated these areas using a cross-sectional approach, that is, they have looked at geographic, socio-economic and health food characteristics of food deserts, as they existed at specific moments in time. Only a small number of studies have conducted longitudinal analysis of food deserts (see for example, Semple & Giguere 2017; McClintock 2008; Larsen & Gilliland 2008). To varying degrees, these studies use historical data to analyze the degree to which non-food desert tracts converted into food desert tracts and then attributed the conversion process to various socio-economic processes such as suburbanization, changing dynamics in the supermarket industry, and changes in spatial patterns of capital flows.

A longitudinal approach to studying food deserts is useful for understanding how historical circumstances and processes underlie the emergence and persistence of this phenomenon. However, it is often desirable to model the trajectory of these food insecure areas in order to predict their future geographic size and likely impacts on the local economy. Information derived from such modeling exercises is useful to planners and local policy makers concerned with community development and improving the overall economic base and quality of life of local communities. This chapter reports on a Markov Chain model that was developed to study the likelihood of census tracts converting to either food desert census tracts or non-food desert census tract over the next two decades in Ypsilanti, Michigan.

2. THE STUDY AREA

Ypsilanti, Michigan is a small urban center located approximately 35 miles southwest of Detroit, and about 15 minutes south east of Ann Arbor.

The study area is centered on the City of Ypsilanti, but it also includes areas in the adjacent townships of Ypsilanti and Pittsfield, which can be considered part of the Ypsilanti urban area. The City of Ypsilanti itself is only about four square miles in area. The total census population in 2010 was 19,500 of whom 60% was white, 31% African-American, 3% Asian, 3% Hispanic, and 3% other ethnicity (United States Census Bureau / American FactFinder 2010).

Historically, the City of Ypsilanti developed as a service center to the surrounding agricultural communities (City of Ypsilanti n.d). It also had a strong auto-manufacturing base dating back to the 1920s (Mann 2003). During the Second World War, both the City of Ypsilanti and the adjacent townships benefitted economically from the presence of the Willow Run aircraft manufacturing plant, which was located in Ypsilanti Township. At its peak, the plant employed about 50,000 people and produced a B-24 bomber at the rate of 1 per hour (Ypsilanti Historical Society n.d.). Many people who worked in the plant lived in the Ypsilanti area and contributed to a robust local economy.

Over the years, the City of Ypsilanti and parts of Ypsilanti Township have experienced gradual loss of their auto-manufacturing base. Both General Motor and Ford motor companies have closed operations along with numerous small industrial and commercial businesses that had upstream and downstream linkages with the automotive giants. This has led to high unemployment rates, outmigration, blighted or abandoned structures, and the emergence of food deserts in some parts of the old industrial area. Census data reveal that between 1970 and 2010 the population of the City of Ypsilanti's fell from 29,538 to 19,435, a decline of 34% (Minnesota Population Center 2011).

Notwithstanding the presence of moderately blighted areas and abandoned buildings in the old industrial core of Ypsilanti, the suburban parts of this small city is sustained by a modern service-based economy, which it shares with the nearby City of Ann Arbor and the adjacent townships. In particular, the presence of Eastern Michigan University with enrollment of 22,000 students has turned Ypsilanti into a 'college town.'

In many ways, Ypsilanti is still slowly transitioning from its old industrial base to one based on higher education and personal services. The pace at which the transition has occurred has been slow, but prospects for growth appears tied to spill over from the more robust economy of the neighboring City of Ann Arbor. This is evidenced, for example, by new strip development that is occurring along Washtenaw Avenue, which provides one of the two important connections between the downtowns of the cities of Ypsilanti and Ann Arbor. The physical development plans that exist for the area foresee increased economic interaction between the two cities and prospects for growth for Ypsilanti (City of Ann Arbor et al. 2009).

In a recent paper by Semple and Giguere (2017) the authors concluded that one or more census tracts that can be designated as "food desert" census tracts have always existed in Ypsilanti over the last four decades. Furthermore, there has been a gradual increase in the number of food desert tracts over the years as the city continued to lose its manufacturing base. The authors defined food deserts as "areas that had grocery density of less than one food store per square kilometer and which were also in the highest quintile of families below the poverty line" (Semple and Giguere 2017 p. 6). The census tracts that coincided with these areas were labeled as food desert census tracts.

Although the food desert census tracts are located in low-income areas of the city that supermarkets have long abandoned, there is uncertainty in any inter-censual period as to which tracts would emerge as food deserts. This is because some of the factors that determine whether a non-food desert census tract converts into a food desert census tract are hard to predict. For example, relocation decisions of factories or other businesses could affect neighborhood median income and hence the determination as to whether a census tract is a 'food desert' or not. Similarly, relocation decisions of markets or supermarkets could affect food store density in a census tract and hence the determination as to whether a tract is labeled a food desert or not. However, it is this randomness that allows the pattern of food deserts in the area over time to be modeled as a Markov process. While the recent research by Semple and Giguere (2017) provided insights into the evolution of food deserts in the Ypsilanti area, it is not clear from that analysis how food deserts in the area are likely to evolve over the next decade or so. This study seeks to provide some insights into this issue.

3. Markov Chains

A Markov Chain is a method for modeling stochastic processes, that is, processes in which there is uncertainty or randomness in the values that variables may assume during their observation time spans (Beichelt 2006 p. 93). These observations may be either continuous or at discrete moments in time. A Markov Chain predicts future scenarios by calculating the likely sequence of states that a system adopts as it moves through time. In the context of this paper, "system" refers to census tracts in Ypsilanti with respect to their food desert status. At any point in time, census tracts can be in either of two states with respect to their food desert status, that is, they can either be food deserts or non-food deserts. According to the Markov assumption, the probability of a tract assuming a given food desert state at a given point in time depends only on the immediate previous state of the system and not on the entire history of the system. Formally, the model can be written in the form of an equation as shown below:

$$\Pr(X_{t+1} = i_{t+1} \mid X_t = i_t, X_{t-1} = i_{t-1}, \ldots, X_1 = i_1, X_0 = i_0) = \Pr(X_{t+1} = i_{t+1} \mid X_t = i_t) \quad (1)$$

Basically, the equation states that the probability distribution of the state of a system at time t+1, given a series of past states of the system is the same as the probability distribution of the state of the system at time t + 1, given only the last state of the system, i.e., the state of the system at time t.

Figure 3.1 is a diagrammatic representation of the simple two-state Markov Chain model used in this analysis. In any given time period, a census tract can be either a food desert census tract (FD) or a non-food desert census tract (NFD). The α and β symbols represent the probabilities of each tract remaining in their particular state during a given period (i.e., transitioning to the same state). Similarly, 1- α represent the probability of a non-food desert tract transitioning to a food desert tract during the period while 1–β represents the probability of a food desert tract transitioning to a non-food desert tract during the period. The analytical task is to estimate these initial transition probabilities from empirical data and then use the probabilities to predict the state of the system' at different time intervals in the future. At each observation time span, the current set of probabilities is used as input to predict the next set of probabilities; hence the metaphor of a "chain" to describe the Markov process.

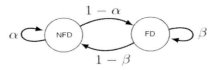

Figure 3.1. Two-state Markov Model.

The various transition values in the Figure 3.1, are better represented in matrix format, where they can be extensively manipulated.

$$P = \begin{bmatrix} \alpha & 1-\alpha \\ 1-\beta & \beta \end{bmatrix}$$

The matrix showing the initial probabilities is called the *transition probability matrix* and is estimated from empirical observations. Once estimated, the transition probability matrix is used to predict the likelihood of census tracts moving from state i to j in n future steps. Based on the Kolmogorov-Chapman equations in mathematics, the predictive process can be reduced to simply multiplying the transition probability matrix by itself n times, where n represents the number of discrete time steps into the future for which the prediction is being made (*Ross 2010* p. 195-196). After a number of steps into the future, the predicted probabilities reach an equilibrium or steady state where additional steps only result in almost identical predicted probabilities. At this point, the transition probability matrix is said to have reached its limit (*Ross 2010* p. 214-216). Once it has reached its limit, a Markov Chain can no longer be used for prediction because the current matrix has become independent of the transition probability matrix.

In this study, only a first-order Markov process was considered. This means that the system only has memory of its immediate previous state. If no memory exists of its immediate previous state, the process is not Markovian and the initial transition probability matrix cannot be used for prediction. Consequently, a statistical test of independence is usually applied to the empirical data to test for this Markovian property before predictions are done with the transition matrix. The null hypothesis must be rejected for the process to be characterized as Markovian.

Markov Chains have long been used in geography to model sequential changes in land uses and other phenomena (Clark 1965; Bell 1974; Bell & Hinojosa 1977; Muller & Middleton 1994; McMillen & McDonald 1991; Weng 2002; Ahmed & Ahmed 2012; Iacono et al. 2015).

4. Methodology

4.1. Data

Data on the number and location of food deserts in Ypsilanti were assembled as part of a food desert study carried out by Semple and Giguere (2017) for Ypsilanti, Michigan. That study identified food desert and non-food desert census tracts for the Ypsilanti area for each of the four inter-censual periods between 1970-2010. Data to create factor maps used for visualizing the Markov process were obtained from the US census. Such data included the number of whites and blacks by census tracts and median income by census tracts were assembled. Data on food store density by census tracts were obtained from datasets available to this author from the study by Semple and Giguere (2017).

Since the goal of the study was to make predictions for 2020 and 2030, the census period 1990-2000 was used as the base period for calculating the transition matrix. Selecting 1990-2000 as the base period provided an opportunity to validate the predicted Markov food deserts with food deserts computed from actual 2010 census data. Being able to validate the model against 2010 data increased confidence in its ability to predict for 2020 and 2030.

Food desert vector maps for 1990 and 2000 were converted into raster format and imported into the TerrSet software. For each pixel in the map, the Markov routine in the software examined the labels of the pixel in the 1990 layer and again in the 2000 layer and kept tract of those that transitioned from 'non-food desert' in 1990 into non-food desert in 2000 (i.e., remained as non-food desert), and those that transitioned from non-food desert into food desert. It also counted the number of times pixels labeled as "food desert" in 1990 transitioned into food desert and non-food desert pixels in 2000. A frequency table of the various transitions was produced and used to compute a transition probability matrix (Table 3.4.1). The transition probabilities in each row of the matrix sum to 1 to account for all the transitions between states j for a particular class, i.

Table 3.4.1. Transition Probability Matrix

	1990	2000 NFD	2000 FD
P =	NFD	0.8571	0.1429
	FD	0.6000	0.4000

Once the transition probability matrix was computed, a statistical test of independence was carried out to test for the Markov property in the study dataset. Specifically, the test sought to determine whether the 'states' of census tracts with respect to food deserts in 1990 were related to the 'states' of census tracts with respect food deserts in 2010. Following Weng (2002) the test of independence was done by comparing the observed transitions for each state of the system between 1990 and 2010 with the expected transitions for the same period under the Markov assumptions and noting whether they were independent of each other. The test statistic, K^2, is shown below:

$$K^2 = \sum_i \sum_k (O_{ik} - E_{ik})^2 / E_{ik} \qquad (2)$$

where O_{ik} = the observed number of raster cells that transitioned from states i to j between 1990 and 2010, and
E_{ik} = the expected number of raster cell transitions based on the Markov property.

Again, following (Weng 2002) the expected number of transitions in the test statistic, E_{ik}, was computed on the basis of the Chapman-Kolmogorov equations. According to the equations, the probability of transitioning from state i to k can be obtained by first computing the transition from state i to j, an intermediate state, and then from j to k, the final state (Ross 2010). Thus, we can compute the number of cells that transitioned from state i to j between 1990 and 2000 and then multiply those by the number of cells that transitioned between 2000 and 2010. The result is then scaled by the total number of cells involved in the transitions. The formula is:

$$E_{ik} = \sum (E_{ij})(E_{jk}) / E_j \qquad (3)$$

where E_{ij} = number of cells that transitioned from state i to j during the period 1990 to 2000,
E_{jk} = number of cells that transitions from state j to k during the period 2000 to 2010,
E_j = the number of raster cells in category j in the study area in 2000.

Since the K^2 test statistic follows a Chi-Square distribution, it was tested for statistical significance against the theoretical Pearson Chi Square distribution with $(M-1)^2$ degrees of freedom, where M is the number of states in the system. The critical region was set at alpha = 0.05 with 1 degree of freedom. Thus, any computed K^2 value greater than 3.841 led to the conclusion that the null hypothesis must be rejected. The computed K^2 was 22.02, which meant that the hypothesis of statistical independence was rejected; therefore first-order Markovian dependence was established in the dataset. A test for stationarity was also done to ensure that the study prediction period occurred before the system reached its steady state. This was done using the markovchain library in R. Our prediction period occurred well before steady state.

Once Markov dependence was established in the dataset, the transition matrix was used for predicting the probabilities of census tracts transitioning from state i to j in future time steps. The likelihood of a census tract transitioning from state i to j in n steps was calculated using the formula:

$$P_{ij}^{(n)} = P_i * P_{ij}^n \tag{4}$$

where $P_{ij}^{(n)}$ = the predicted probabilities;
$\quad P_i$ = the initial probability distribution, or it could be a vector of states; and
$\quad P_{ij}^n$ = the transition probability matrix raised to the power n.

The predicted probabilities after the first step are, by definition, just the initial probabilities. They can also be calculated from the formula by multiplying the initial probabilities by the transition matrix raised to the 0 power. For our study, we were interested in predicting Markov probabilities for the years 2020 and 2030, i.e., for two and three steps beyond the initial starting point.

4.2. Combining Markov Chain with Cellular Automata

One of the problems with Markov modeling is that the predicted probabilities do not have a spatial dimension so the results are displayed as tables and not as maps. However, geographers have experimented with ways for cartographically visualizing Markov predictions. Indeed, the TerrSet software is able to combine the Markov Chains with stochastic cellular automata (CA) to create prediction maps. Many researchers have used this approach for Markov visualization (Ahmed and Ahmed 2012).

A cellular automaton (CA) is a cell-based or grid structure, similar to a raster, in which a simple set of rules are applied to the current state of the system to predict the next future state of the system. At each time step, each cell in the grid is set to a value that reflects whether or not certain rules or conditions were fulfilled at the cell's location as well as neighboring cell locations. The rules control how the system evolves and are reapplied over many time steps to simulate the growth pattern of the system. In a GIS, the cellular automaton rules are applied against one or more input base maps. When a Markov Chain model is linked to a cellular automata model for visualization of the Markov transitions, the cellular automata model provides the input map layers as well as rules against which conditions can be tested. The Markov model supplies the transition probabilities, which controls the likelihood of cells transitioning from one state to another.

In this study, the TerrSet software was used to generate Cellular Automata-Markov prediction maps for 2010, 2020, and 2030. The maps show the predicted state of each pixel at future time periods. The critical inputs for creating the prediction maps were the food desert map for the year 2000, the transition areas matrix, and a set of suitability layers. Only the suitability layers have not been previously discussed.

Suitability layers for CA-Markov prediction in TerrSet are the base layers against which the cellular automaton model applies transition rules. These layers provide geographic expressions to the Markov predictions by supplying pixels that show the most likely location of each Markov class at different time steps into the future. Since our model dealt with two classes, food deserts and non-food deserts, two suitability layers were required. Each suitability layer was created by combining a set of weighted input layers called *factor maps*. The factor maps used to create the Markov food desert suitability layer were census tract layers showing low median income, low food store density, high African American population and low white population. Weights were attached to each layer to reflect their relative importance. Similarly, census tracts layers showing high median income, high food store density, high white population, and low African American population were used as the factors to create the Markov non-food desert suitability layer.

The values in the factor maps were standardized using fuzzy logic but re-scaled to 0 - 255. Weights attached to the different factor maps were subjectively determined. The Multi Criteria Evaluation (MCE) tool in TerrSet was used to combine the factor maps into suitability layers, one for each Markov class. The two suitability layers were then grouped into a TerrSet group layer and fed into the CA-Markov routine along with the food desert file for 2000, the transition probability matrix, the number of iterations that are required, and the default 5 x 5 filter for the cellular automata. The software evaluated the input variables and produced prediction maps for the specified periods, i.e., 2010, 2020 and 2030.

The simulated food desert map for 2010 was compared with the reference food desert map for 2010 to validate the CA-Markov visualization. This was done using the validation tools in TerrSet. The software calculates the well-known Kappa Index of Agreement (KIA), several modified Kappas, and measures of disagreements and agreements (Pontius & Millones, 2011).

5. RESULTS

The predicted transition probabilities for the decades ending 2020 and 2030 are shown in Table 3.5.2. The table shows that the probability of a census tract which started as a non-food desert tract in 2010 and remaining as a non-food desert in 2020 is predicted to be 0.81091. It will marginally decline to 0.8085 by 2030. In the initial 1990/2000 transition matrix, the probability of non-food desert tracts remaining as non-food desert tracts was 0.8571. The distinctive pattern is thus moderate decreases in the likelihood of non-food desert tracts remaining as non-food deserts tracts between 2000 and 2010 (due to increases in food desert tracts), followed by marginal decreases between 2010 and 2030.

The probability of a census tract transitioning from non-food desert to food desert status was 0.1429 in the original transition matrix and is predicted to increase to 0.18909 by 2020 and 0.19151 by 2030. Again, these data point to a period of moderate increases in food desert tracts between 2000 and 2010, followed by marginal increases onwards to 2030. However, there is a high probability that a food desert tract could resort to a non-food desert tract. Indeed, the historical food desert maps presented in Semple & Giguere (2017 p.7) show evidence of this reconversion.

Table 3.5.2. Predicted Probabilities for Food Desert and Non-Food Desert Census Tracts

		2020	
2010		NFD	FD
	NFD	0.81091	0.18909
	FD	0.79392	0.20608

		2030	
2020		NFD	FD
	NFD	0.80849	0.19151
	FD	0.80412	0.19588

The CA-Markov maps are shown in Figures 3.2 and 3.3. Figure 3.2 shows that the spatial pattern of food deserts on the reference map for 2010 is somewhat different from the pattern on the simulated map for 2010. This is mainly because the factor maps that were used to predict the location of the 2010 transitions were based on patterns that existed in 2000 and did not include the sudden emergence of food deserts in the southeastern quadrant of the study area in the period after 2000. In that area, the closing of two supermarkets in the Gault Village mall contributed to problems of food accessibility in that area.

Figure 3.3 shows simulated food deserts for 2020 and 2030 based on the 1990-2000 transition matrix. As expected, compared to the moderate increases in food deserts in 2010, the 2020 and 2030 maps show only marginal increases in food desert expansion, which corroborates the probabilities depicted in the Table 3.5.2.

Simulating the Future of Food Deserts in Ypsilanti ...

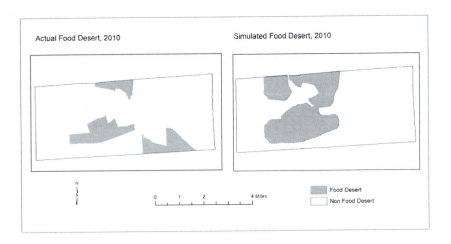

Figure 3.2. Actual and Simulated Food Deserts for 2010 based on 1990-2000 Transition Matrix.

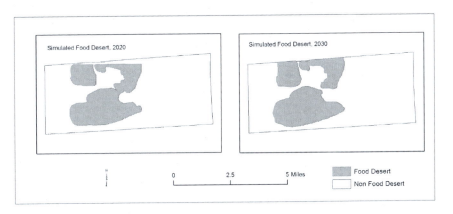

Figure 3.3. Simulated Food Deserts, 2020 and 2030 based on 1990-2000 Transition Matrix.

5.1. Model Validation

Table 3.5.3 shows various Kappa indices computed by TerrSet for the model validation. The traditional Kappa Index of Agreement is denoted as Kstandard, and was calculated at 0.5732. This suggests an overall moderate to good agreement between the predicted food desert map and the actual food desert map for 2010. The Kno statistic is a general statistic similar to Kstandard, but is modified to adjust for large quantity errors (TerrSet Help 2015). It is interpreted similar to the traditional KIA, so the Kno statistic of 0.6940 reflects an even better overall agreement between the maps. Klocation denotes Kappa for grid cell or spatial agreement while KlocationStrata is a Kappa for class-level agreements (TerrSet Help 2015). Both measures indicate good, though not excellent, overall agreement of pixels at the location and the class levels for the two maps. Overall, the validation results suggest that the CA-Markov map moderately reflected the food desert dynamics in the study area.

Table 3.5.3. Various Kappa Indexes

Kstandard	0.5732
Kno	0.6940
Klocation	0.7497
KlocationStrata	0.7497

DISCUSSION

The period 1970-2010 was marked by an increasing trend in food deserts in the Ypsilanti Region (Semple and Giguere 2017). Assuming current economic conditions in the Ypsilanti region remain as they are for the next decade and a half, then the Markov predictions of very slight increases in food deserts appear plausible. However, a recently released economic forecast for Washtenaw County, in which the Ypsilanti region is located, predicts sustained job growth, growth in real wages, increases in employment levels, and only moderate inflation between 2017 and 2019 (Ehrlich et al. 2017). Under these assumptions, there might be a slowdown in the rate at which food deserts appear in the Ypsilanti region. In fact, some areas may revert back to non-food desert status.

The CA-Markov visualization of food desert expansion appears plausible. However, the TerrSet Help notes that during the prediction process, the cellular automata contiguity filter gives significantly more weight to pixels that are close to a land cover class compared to pixels that are further away. Thus, predicted areas are more likely to be located closer to existing areas of a class. This effect is seen on the maps where the predicted food desert areas for 2010, 2020 and 2030 simply grow outwards from existing areas. In the past, census tracts that met the requirements for food deserts have not always been located in contiguous census tracts, so the CA-Markov is not fully equipped to capture these more dynamic patterns, particularly if the prediction is outside the range of the patterns in the input factor maps. Still, the method provides a reasonable approximation of where newer food desert areas may appear in the city.

One specific weakness of the model is that it relies on a transition probability matrix that was constructed with 1990-2000 data. While it would have been better to use a transition probability matrix with 2000-2010 cell transitions, this would have prevented model validation since 2010 is the latest date for which actual food desert data are available. Also, food desert data were not available for the intervening period 2000 to 2010 because food deserts were calculated using census data. The major effect of using the 1990/2000 transition probability matrix is that it slightly under predicts the extent of food deserts in the Ypsilanti region because the emergent food deserts in the southwest quadrant were not taken into consideration.

A more general weakness of the model is that the probabilities in the transition matrix are fixed, i.e., they are stationary or time homogenous. This means that predictions made with these probabilities do not take changing economic circumstances into consideration. Thus, the longer the time period over which the predictions are made, the less reliable the model.

Given the slight predicted increases in food desert areas during the period 2020 to 2030, an important question is what policy initiatives are available to local policy makers for reversing the expansion or reducing the extent of these areas. There is a growing body of literature on this topic based on the experiences of various communities. Strategies that have been utilized include subsidizing companies to build new food stores or renovate existing ones in food desert areas (Ungar-Sargon 2016); the use of zoning regulations to prevent further establishment of liquor stores and fast food restaurants in food desert areas (Centers for Disease Control and Prevention 2012); promotion of urban food gardens (Cyzman, Wierenga & Sielawa 2009; Twiss et al. 2003; see Beckford & Igbokwe this volume); encouraging convenience stores to carry larger supplies of fresh fruits and vegetables (Dannefer, et al. 2012; Ramos, et al. 2015); and improving bus routes in low income areas to help people to easily access existing supermarkets (McCann 2006).

From a review of available literature, it appears that singular implementation of the types of initiatives mentioned above are not highly successful in reversing food deserts (Abraham 2016; Reisig & Hobbiss 2000). Instead, there is a growing realization that food deserts are symptomatic of a range of integrated socioeconomic issues and their reversal requires a holistic approach that simultaneously addresses various community issues such as low wages, lack of education, access to reasonably priced food, and knowledge about nutrition, food preparation, and their relationship to health.

Conclusion

Markov Chains have been widely used for stochastic predictions. In this study, it has been shown that Markov Chains can be used to predict the likely long-term pattern of food deserts in the Ypsilanti area. The model is simple to use and can provide planners with an effective tool to forecast geographic expansions in food deserts.

The Markov model developed for this study predicted moderate expansion in food deserts for the Ypsilanti region between 2000 and 2010 followed by marginal increases from 2010 onwards to 2030. However, it appears that food deserts will remain a feature of the urban landscape of Ypsilanti for the near future. Assuming reliability in the probabilities, policy makers can use this type of information to guide local economic development in the direction of reducing expansion in food deserts.

REFERENCES

Abraham, T 2016, 'Why food deserts require more than just supermarkets,' *Generosity*, Available from: http://generocity.org/philly/2016/02/29/uplift-solutions-food-deserts/ (January 2, 2018).

Ahmed, B & Ahmed, R 2012, 'Modeling Urban Land Cover Growth Dynamics Using Multi-Temporal Satellite Images: A Case Study of Dhaka, Bangladesh,' *ISPRS International Journal of Geo-Information*, 1(3), pp.3-31.

Alameda County Public Health Department 2008, 'Food access and liquor Stores. Life and Death from Unnatural Causes,' *Health and Social in Alameda County*. Oakland, California. *Alameda County Public Health Department*, pp. 97-101.

Babey, SH, Diamant, AL, Hastert, TA & Harvey, S 2008, 'Designed for Disease: The Link between Local Food Environments and Obesity and Diabetes,' *UCLA Center for Health Policy Research*: Los Angeles, CA, USA, 2008. Available from: *http://www.publichealthadvocacy.org/PDFs/RFEI%20Policy%20Brief_finalweb.pdf* (January 2, 2018).

Beaulac, J, Kristjansson, E, & Cummins, S 2009, 'A Systematic Review of Food Deserts, 1966-2007,' *Preventing Chronic Disease*, 6(3), p. A105.

Beichelt, F 2006, *Stochastic processes in science, engineering, and finance*, Boca Raton: Chapman & Hall/CRC.

Bell, E 1974, 'Markov analysis of land use change - an application of stochastic processes to remotely sensed data,' *Socio-Economic Planning Sciences*, 8(6), pp. 311-316.

Bell, E & Hinojosa, R 1977, 'Markov analysis of land use change: Continuous time and stationary processes,' *Socio-Economic Planning Sciences*, 11(1), pp. 13–17.

Block, J, Scribner, R & DeSalvo, K 2004, 'Fast food, race/ethnicity, and income. A geographic analysis,' *American Journal of Preventive Medicine*, 27(3), pp. 211-217.

Centers for Disease Control and Prevention (CDC) (2012), *Zoning to encourage healthy eating. Available from: URL* http://www.cdc.gov/phlp/winnable/zoning_obesity.html (December 28, 2017).

City of Ann Arbor, Pittsfield Township, City of Ypsilanti, and Ypsilanti Township 2009, Re-Imagining Washtenaw Avenue. A Vision for Corridor Redevelopment. *Report prepared by Washtenaw County and the Washtenaw Avenue Action Team.*

City of Ypsilanti No Date, *'Comprehensive Development Plan,'* Available from: http://www.cityofypsilanti.com/DocumentCenter/View/299 (December 28, 2017).

Clark, WAV 1965, 'Markov Chain Analysis in Geography: An Application to the Movement of Rental Housing Areas,' *Annals of the Association of American Geographers*, 55(2), pp. 351–359.

Cyzman, D, Wierenga J, & Sielawa, J 2009, 'Pioneering Healthier Communities, West Michigan,' *Health Promotion Practice*, 10(2), pp.146S-155S.

Dannefer, R, Williams, D, Baronberg, S, & Silver, L 2012, 'Healthy Bodegas: Increasing and Promoting Healthy Foods at Corner Stores in New York City,' *Am J Public Health* 102(10), pp. e27–e31.

Dutko, P, Ver Ploeg M, & Farrigan, T 2012, 'Characteristics and Influential Factors of Food Deserts,' ERR-140, *US Department of Agriculture, Economic Research Service.*

Ehrlich, G, Fulton, G, Grimes, R & McWilliams, M 2017, 'The Economic Outlook for Washtenaw County in 2017-19,' *University of Michigan.* Available from: http://media.mlive.com/ann-arbor-business_impact/other/A2%202017%20Outlook.pdf (December 28, 2017).

Fraser, L, Edwards, K, Cade, J & Clarke, G 2010, 'The Geography of Fast Food Outlets: A Review,' *International Journal of Environmental Research and Public Health*, 7(5), pp. 2290-2308.

Iacono, M, Levinson, D, El-Geneidy, A & Wasfi, R 2015, 'A Markov chain model of land use change in the Twin Cities, 1958-2005,' *Tema - Journal of Land Use, Mobility and Environment*, 8(6), pp. 311-316.

Larsen, K & Gilliland, J 2008, 'Mapping the evolution of food deserts in a Canadian city: Supermarket accessibility in London, Ontario, 1961–2005,' *International Journal of Health Geographics*, 7(1), p.16.

Mann, JT 2003, *Images in America. Ypsilanti in the 20th century,* Charleston, SC: Arcadia.

McCann B 2006, *Community design for healthy eating: how land use and transportation solutions can help*. Robert Wood Johnson Foundation. Available from: http://www.slideshare.net/GeoAnitia/community-design-for-healthy-eating-how-land-use-and-transportation-solutions-can-help (January 2, 2018).

McClintock, N 2008, *From industrial garden to food desert: unearthing the root structure of urban agriculture in Oakland, California, ISSI Fellows Working Papers*, Institute for the Study of Societal Issues, Berkeley, CA: Dept. of Geography, University of California.

McMillen, DP & McDonald, JF 1991, 'A Markov Chain model of zoning change,' *Journal of Urban Economics*, 30(2), pp. 257–270.

Minnesota Population Center 2011, *National Historical Geographic Information System: Version 2.0*, Minneapolis, MN, University of Minnesota.

Muller, MR & Middleton, J 1994, 'A Markov model of land-use change dynamics in the Niagara Region, Ontario, Canada,' *Landscape Ecology*, 9, pp. 151–157.

Pontius, RG & Millones, M 2011, 'Death to Kappa: birth of quantity disagreement and allocation disagreement for accuracy assessment,' *International Journal of Remote Sensing*, 32(15), pp. 4407–4429.

Ramos, A, Weiss, S, Manon M, Harries, C 2015, *Supporting Healthy Corner Store Development in New Jersey*. Philadelphia, PA: The Food Trust.

Reisig, V & Hobbiss, A 2000, 'Food Deserts and How to Tackle Them: A Study of One City's Approach,' *Health Education Journal*, 59, pp. 137-149.

Ross, S 2010, *Introduction to Probability Models (11th ed.)*. Oxford, England: Academic Press.

Schafft, K, Jensen, E, & Hinrichs, C 2009, 'Food Deserts and Overweight School Children: Evidence from Pennsylvania,' *Rural Sociology*, 74(2), pp. 153–77.

Semple, H & Giguere, A, 2017, 'The Evolution of Food Deserts in a Small Midwestern City: The Case of Ypsilanti, Michigan: 1970 to 2010,' *Journal of Planning Education and Research*. DOI: 10.1177/0739456X1770222.

Smith, C & Morton, LW 2009, 'Rural Food Deserts: Low-income Perspectives on Food Access in Minnesota and Iowa,' *Journal of Nutrition Education and Behavior*, 41(3), pp. 176–187.

TerrSet Help 2015, *TerrSet Geospatial Monitoring and Modeling Software, Worcester, Massachusetts*. Worcester, Massachusetts: Clark Labs at Clark University.

Twiss, J, Dickinson J, Duma S, Kleinman T, Paulsen, H & Riveria L 2003, 'Community gardens: lessons learned from California healthy cities and communities,' *American Journal of Public Health*, 93(9), pp. 1435–438.

Ungar-Sargon, B 2016, 'Have City Subsidies to Supermarkets Made NYC Healthier?,' *CityLimit.org*. Available at: https://citylimits.org/2016/04/05/have-city-subsidies-to-supermarkets-made-nyc-healthier/ (December 28, 2017).

United States Census Bureau/American FactFinder, United States Census 2010, Table B02001. *US Census Bureau*. Available at: http://factfinder2.census.gov (December 28, 2017).

Walker, RE, Keane, CR & Burke, JG 2010, 'Disparities and access to healthy food in the United States: A review of food deserts literature,' *Health & Place*, 16(5), pp. 876–884.

Weng, Q, 2002 'Land use change analysis in the Zhujiang Delta of China using satellite remote sensing, GIS and stochastic modelling,' *Journal of Environmental Management*, 64(3), pp. 273–284.

Ypsilanti Historical Society (No Date), Apex Motor Company - 1920-1922. *Ypsilanti Historical Society*. Available from:
http://www.ypsilantihistoricalsociety.org/history/page999993.html (December 28, 2017).

In: Agriculture, Food, and Food Security
Editor: Clinton Lloyd Beckford
ISBN: 978-1-53613-483-4
© 2018 Nova Science Publishers, Inc.

Chapter 4

INFLUENCING FACTORS ON THE IODINE CONTENT OF FOOD: A REVIEW OF ANIMAL SOURCES OF IODINE

Gerhard Flachowsky, Ulrich Meyer, Ingrid Halle and Andreas Berk*
Institute of Animal Nutrition, Friedrich-Loeffler-Institute, Braunschweig, Germany

1. INTRODUCTION

In Chapter Two, the iodine requirements of humans and animals, as well as the iodine content of water and food of plant origin were introduced and discussed. In this chapter, we review the factors that influence iodine content in food of animal origin.

In this regard, we focus on milk, eggs and fish, which are very important iodine sources for human beings, but we also mention the iodine content of meat. In food of animal origin, there is a certain potential for increasing the

---

* Corresponding author email: Gerhard.Flachowsky@fli.de.

concentration of micronutrients relevant to human nutrition including iodine (Rooke, Flockhart & Sparks 2010). On the other hand, there is also the risk of exceeding the recommended upper levels of several nutrients including iodine after special feeding of animals and subsequent high consumption of such biofortified foods by humans (see Joint FAO/WHO Expert Committee on Food Additives (JECFA) (JECFA 1989)). Therefore, it is important to take note of some critical influencing factors on the iodine content of food of animal origin.

We start with a discussion of critical attributes of milk concerning iodine before similar discussions of eggs, meat and finally fish. The chapter ends with some final thoughts.

2. MILK

Animal milk, marine fish (Chilean Iodine Educational Bureau (CIEB) 1952) and eggs are among the most important sources of iodine for humans, particularly in childhood (Chacón Villanueva & Agency 2016; European Food Safety Authority (EFSA) 2013a; EFSA 2013b; Henderson & Gregory 2002; Johner et al. 2012; Johner et al. 2013; Schöne et al. 2009; Flachowsky et al. 2013). This recent research confirms what Broadhead, Pearson & Wilson (1965) established more than 50 years ago. Table 4.2.1 summarizes the iodine content of cow's milk- which is the most important milk for human nutrition in terms of consumption- as presented by various authors for different countries.

For human milk, an average of 51 µg iodine/l and a range between 5 and 90 µg iodine/L are given by Souci, Fachmann & Kraut (2008). The European Food Safety Authority (EFSA 2005; EFSA 2013a; EFSA 2013b) has also reviewed the effects of various I-supplementations to food producing animals and the iodine content of food of animal origin and provided similar results.

Table 4.2.1: Mean iodine content in fresh milk weight samples (µg/L) by various authors

Origin/Country	Mean	Range	Authors
Chile	47	35 – 56	CIEB (1952)
Finland	169		Varo et al. (1982)
UK	230	50 – 550	Wenlock et al. (1982)
USA	200	± 80	Pennington et al. (1995)
England	311	90 – 430	Committee on toxicity of chemicals in food (COT) (2000b)
France	113	84 – 428	French Agency for Food (ANSES) (2005)
Switzerland, 2000 – 2001	124	59 – 199	Haldimann et al. (2005)
Czech Rep.	324		Kursa et al. (2005)
Czech Rep.	489	387 – 601	Travnicek et al. (2006a)
Austria	76	45 – 92	Rysava, Kubackova & Stransky (2007)
Poland	90	86 – 93	Rysava, Kubackova & Stransky (2007)
Switzerland	90	76 – 106	Rysava, Kubackova & Stransky (2007)
Germany	130	93 – 159	Rysava, Kubackova & Stransky (2007)
Belgium	158		Rysava, Kubackova & Stransky (2007)
Slovakia	240	180 – 310	Rysava, Kubackova & Stransky (2007)
Slovakia	325	305 – 345	Rysava, Kubackova & Stransky (2007)
Czech Rep.	471		Rysava, Kubackova & Stransky (2007)
France	207	192 – 221	Rysava, Kubackova & Stransky (2007)
Germany	41	21 – 110	Souci, Fachmann & Kraut (2008)
Iceland	259		Soriguer et al. (2011)
Norway	190		Haug et al. (2012)
Denmark	243		Haug et al. (2012)
Finland	170		Haug et al. (2012)
Sweden	140		Haug et al. (2012)
Iceland	112		Haug et al. (2012)
Spain	198		Arrizabalaga et al. (2015)
Catalonia	146	45 – 204	Chacón Villanueva & Agency (2016)
Germany	105		Schöne, Sporl & Leiterer (2017)

2.1. Influencing Factors on Iodine Content of Milk

2.1.1. Iodine Intake of Cows

Iodine intake of animals is one of the most important influencing factors on the iodine content of milk. Franke (2009) summarized previous studies on the influence of various levels of iodine intake on iodine concentration of milk (see Table 4.2.2 & Figure 4.1) A linear increase in the iodine concentration of milk with increasing iodine intake of the cows was observed and supported by other studies (see Battaglia et al. 2009; Castro et al. 2011; Dahl et al. 2003; Dahl et al. 2004; Hemken et al. 1972; Herzig et al. 1999; Hillman & Curtis 1980; Miller & Swanson 1973; Moschini et al. 2009; Swanson et al. 1990). However, non-linear response curves were calculated by another group of authors (Norouzian 2009; Norouzian et al. 2011). A very high milk iodine concentration (>2500 µg/kg) was measured by Schöne et al. (2009) after supplementation of dairy cow diets with 10 mg iodine/kg dry matter (DM).

In a study of 200 farms Castro et al. (2011) grouped 30 farms with a low iodine concentration (1.20 mg/kg DM) and 30 farms with a high iodine concentration in the feed (1.81 mg/kg DM) and measured 103 and 554 µg iodine/kg milk in 2007 as well as 146 and 487 µg iodine/kg milk in 2008 for the low and high supplementation respectively. The large differences in the iodine-milk content are surprising because of the small differences in the I-concentration of feed. Other factors, such as the presence of goitrogens (see 2.1.3) or farm management (see 2.1.4.2) may also have influenced the concentration of iodine in milk.

2.1.2. Iodine Species

According to the European Union Legislation, sodium iodide (NaI), potassium (KI) iodide, calcium iodate hexahydrate (Ca(IO$_3$)$_2$ x 6H$_2$0 as well as calcium iodate (Ca(IO$_3$)$_2$ anhydrous, are approved as nutritive additives for feed supplementation in the European Union (EU) (EU 2015).

The effect of different iodine sources in dairy cattle feed on the transfer into milk has been the subject of some research (see Franke et al. 2009a; Franke et al. 2009b). Several studies comparing both iodine sources showed no clear differences in the iodine concentrations of milk after daily application (Bretthauer, Mullen & Moghissi 1972; Lengemann 1969). In contrast, Leskova (1969) described higher milk iodine concentrations as well as longer excretion time following oral application of potassium iodate compared to potassium iodide. Franke et al. (2009b) detected similar tendencies for higher iodine concentrations in milk when applying iodate compared to iodide (see Table 4.2.2 & Figure 4.1).

It is suggested that iodate may lead to higher iodine concentrations in milk, since for iodide, higher storage losses are expected due to its instability in the presence of oxygen, excessive aeration, sunlight, and ultraviolet light (Diosady et al. 1998; Waszkowiak & Szymandera-Buszka 2008).

2.1.3. Goitrogens in the Rations

Goitrogens are agents that may cause thyroid enlargement by interference with the thyroid hormone synthesis and secretion, including feed-back mechanisms of thyroid stimulating hormone (TSH) and TSH releasing factor. They either influence the iodine uptake into the thyroid, the oxidation of iodide to elemental iodine with the subsequent transfer into the thyroglobulin, the synthesis of thyroid hormones, or the proteolysis or release of the thyroid hormones (Gaitan 1990; Tripathi & Mishra 2007). Plants that may contain goitrogen substances include plants of the cruciferous family, inclusively rape and kale, as well as raw soybean, beet pulp, millet, linseed, cyanogenic strains of white clover and sweet potato (Castro et al. 2011).

Table 4.2.2: Influence of iodine source potassium iodide and calcium iodate hexahydrate on iodine concentration of milk (µg per liter)

Supplementation (mg iodine/kg DM)	Unsupplemented control (0.2 mg iodine/kg DM)	0.5	1	2	3	4	5
Potassium iodide	83	158	214	550	638	1085	1464
Calcium iodate hexahydrate	72	188	231	584	930	1188	1578

Source: Franke et al. (2009a).

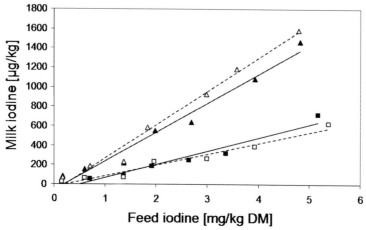

Source: Franke et al. 2009a; Franke et al. 2009b.

Figure 4.1. Dependence of the milk iodine concentration on the kind of iodine supplementation in diets without and with iodine antagonists via rape seed meal (RSM; n = 8, ▲, control/iodide; △, control/iodate; ■, RSM/iodide; □, RSM/iodate.

Glucosinolates diminished the milk iodine concentration by inclusion of rapeseed feeds. The extent of the rapeseed mediated milk iodine decrease does not seem to depend on the amount of glucosinolates ingetsted. At 12 mmol glucosinolates per cow per day there was a milk iodine content reduction of 2/3 in comparison to the control (Franke et al. 2009b) (see Figure 4.1).

The following regression equations were calculated where y is the iodine concentration of the diet (mg/kg DM) and x is the iodine concentration in milk (μg/kg)):

Control (diet without rapeseed meal, added potassium iodide):

$y = 342.2 x - 73.1$; ($R^2 = 0.98$),

Control (diet without rapeseed meal, added calcium iodate):

$y = 298.3 x - 64.0$; ($R^2 = 0.97$),

With glucosinolates (diet with rapeseed meal, added potassium iodide):

$y = 136.5 x - 67.1; (R^2 = 0.94)$,

With glucosinolates (diet with rapeseed meal, added calcium iodate):

$y = 112.0 x - 24.3; (R^2 = 0.96)$.

In the study of Franke et al. (2009b) the milk iodine depression was observed at various iodine supplementations. Since the percentage of milk iodine reduction by rapeseed stayed constant, it was shown that the inhibitory effect of the glucosinolates would not be impeded by higher iodine dosages. Apart from rapeseed products, other cruciferous plants such as *Crambe abessynica* and other plants with glucosinolates, show similar effects on the iodine concentration of milk (Schöne et al. 2006a). About 0.8 and 1.45 kg DM of crambe cake or crambe meal reduced the iodine concentration of milk from 226 on average to about 120 and 97 µg/L milk (Table 4.2.3).

Table 4.2.3: Influence of various amounts of crambe cake (50.4) and crambe meal (77.4 mmol glucosinolates/kg DM) in rations of dairy cows on the iodine concentration of milk (n = 10 cows; 0.8 mg iodine/kg DM)

Crambe in concentrate (%)	Crambe cake intake (kg/d)	Glucosinolate intake (mmol/d)	Iodine in milk (µg /L)	Crambe meal intake (kg/d)	Glucosinolate intake (mmol/d)	Iodine in milk (µg/L)
0	0	1.0	271 ± 64	0	1.1	182 ± 30
15	0.77	39.0	142 ± 40	0.83	59.8	95 ± 28
30	1.35	75.2	117 ± 33	1.53	114.7	77 ± 14

Source: Böhme et al. (2005).

2.1.4. Further Influencing Factors

2.1.4.1. Season

In many studies, most milk samples from indoor/winter keeping, show a higher iodine content than milk during outdoor/summer feeding, probably caused by a higher mineral supplementation and/or a lower content of antinutritive substances (e.g., glucosinolates) in the winter diet (see Table 4.2.4).

Table 4.2.4: Influence of summer (outdoor, grazing) and winter (indoor) animal feeding and keeping on the iodine concentration of bulk milk (µg/kg) in some European studies

Author(s)	Country	Type of animal feeding/keeping		Remarks
		Outdoor (Summer)	Indoor (Winter)	
Broadhead, Pearson & Wilson (1965)	UK	10 – 60	70 – 210	
Lee et al. (1994)	UK	90	210	
Fiedlerova (1998)	Czech Republic	119	205	
Dahl et al. (2003)	Norway	88	232	Conventional
Dahl et al. (2003)	Norway	60	127	Organic
COT (2000b)	England	90	210	1991/92
		200	430	1998/99
ANSES (2005)	Spain	62	203	
Hejtmankova et al. (2006)	Czech Republic	212	251	
Travnicek et al. (2006a)	Czech Republic	351	494	
Paulikova et al. (2008)	Slovakia	155	127	Cow's milk
Paulikova et al. (2008)	Slovakia	56	198	Sheep's milk
Paulikova et al. (2008)	Slovakia	48	89	Goat's milk
Brzóska, Szybiński & Śliwiński (2009)	Poland	100	147	

Author(s)	Country	Type of animal feeding/keeping		Remarks
		Outdoor (Summer)	Indoor (Winter)	
Hampel, Kairies & Below (2009)	Germany	108	134	
Soriguer et al. (2011)	Iceland	247	270	
Rozenska et al. (2011)	Czech Republic	38	72	Sheep's milk
Soriguer et al. (2011)	Spain	247	270	
Haug et al. (2012)	Norway	92	122	
Rey-Crespo, Miranda & Lopez-Alonso (2013)	Spain	35	73	Organic farming
Johner et al. (2012)	Germany	87	110	
Crnkić et al. (2015)	Bosnia-Herzegovina	51 (18 – 104)	84 (24 – 216)	Values from 5 regions
Haug et al. (2015)	Norway	79	111	
Troan et al. (2015)	Norway	128 (±92)	212 (±78)	

2.1.4.2. Farm Management

Milk from organic farming contained less iodine compared with milk from conventional farms (Table 4.2.5). The differences in iodine content between organic and conventional milk can mainly be explained by the variation in feeding practices (Crnkić et al. 2015; Srednicka-Tober et al. 2016). A reduced use of iodine-containing mineral mixtures and also less frequent teat-dipping (see 2.1.4.3) on organic farms could explain the lower iodine content in "organic" milk.

2.1.4.3. Teat-Dipping with Iodine Containing Substances

Teat dipping with iodine-containing solutions is still common in many countries for teat disinfection and preventing transmission of contagious mastitis pathogens from cow to cow (Flachowsky et al. 2013). Today, in some countries, iodine is replaced by other substances (e.g., chlorhexidine, glycerine, panthenol etc.) for the dipping operation.

Table 4.2.5: Influence of type of farming on the iodine concentration of bulk milk (µg/kg) in some European studies

Author(s)	Country	Type of farming		Remarks
		Organic	Conventional	
Rasmussen, Larsen & Ovesen (2000)	Denmark	167 196 139	268 306 235	 Winter Summer
Dahl et al. (2003)	Norway	93	231	Winter
Dahl et al. (2003)	Norway	51	167	Summer
Jahreis, Leiterer & Fechner (2007)	Germany	112	169	
Rozenska et al. (2011)	Czech Republic	302	350	
Bath, Button & Rayman (2012)	United Kingdom	144	250	
Johner et al. (2012)	Germany	58	112	
Köhler et al. (2012)	Germany	92	143	
Rey-Crespo, Miranda & Lopez-Alonso (2013)	Spain	78	157	
Arrizabalaga et al. (2015)	France	57	198	
Stevenson, Drake & Givens (2018)	United Kingdom	241	427	(Winter and summer; supermarket)

Depending on the iodine content of the disinfectant and on the manner of application of the solution, the iodine concentration of milk could be significantly increased up to between 7 (Berg & Padgitt 1985) and 150 µg I/L milk (Castro et al. 2012). An increase of 409 µg I/L milk when spraying with disinfectants with a high I-content (10 g I/L) after milking was measured (Castro et al. 2012).

Conrad & Hemken (1978) investigated the mode by which the iodine entered the milk. They concluded that the primary reason for increased iodine appears to be the absorption through the skin and entry into the milk by the milk synthesis process rather than by contamination from the surface of teats. On the other hand, Rasmussen, Galton & Petersson (1991) suggested from their studies, that iodine residues in milk originate mainly from contamination of the teat surface rather than from iodine absorption through the skin, because iodine residues in milk were insignificant when teats were cleaned with towels. More details are described by Flachowsky et al. (2007) and Flachowsky et al. (2013).

2.1.4.4. Cattle Breed, Lactational Stage and Small Ruminants

Less information is available about the influence of cattle breed on the iodine content of milk. Franke, Bruhn & Osland (1983) mentioned an influence of breed on milk iodine concentration, but later papers could not confirm this finding. Ranz, Scherer-Herr & Rambeck (2001) showed decreased concentrations of iodine in the milk of some breeds, but Kroupova et al. (1999) found no differences in iodine content of milk after comparing various breeds in their tests.

Research by Battaglia et al. (2009) showed no influence of milk yield on milk iodine concentration. However, Falkenberg et al. (2002) described a negative correlation between iodine concentration of milk, and milk yield of cows. These results are understandable, if the feed intake does not linearly increase with the milk yield. In the case of a linear correlation of dry matter intake with iodine intake and milk yield, the iodine concentration in milk should not be changed.

Information about the effects of iodine supplementation on milk composition in smaller livestock like goats and sheep are rather scarce. Only a few studies with small ruminants are available to assess influencing factors on the iodine content of their milk (Table 4.2.4). Travnicek & Kursa (2001) investigated the milk iodine concentration in ten sheep flocks and found 105 µg/L. The corresponding value for four farms where sheep had access to mineral licks containing 35 mg iodine/kg was 243 µg iodine/L milk. Similar results were measured for goats. Goats and sheep showed a similar influence of outdoor (summer) and indoor (winter) keeping compared with dairy cows (Paulikova et al. 2008; Rozenska et al. 2011). Sheep kept under organic farming conditions showed a lower milk iodine concentration than sheep kept under conventional farming conditions (Rozenska et al. 2011).

Nudda et al. (2009) fed ten goats each of the Sarda population with an unsupplemented hay-concentrate diet containing 0.35 mg iodine/kg DM and supplemented two other groups with 0.45 and 0.90 mg potassium iodide/day. The iodine intake amounted to 0.65; 1.00 and 1.35 mg/d and the mean milk iodine concentrations were 60, 79 and 130 µg/L. The carry-over of iodine from feed into milk varied between 9.4 and 11.3%, without significant differences caused by the amount of iodine supplementation.

In general, goats and sheep showed a similar behaviour concerning carry-over of iodine in the milk compared to dairy cattle. Therefore, the upper level for feeding of lactating minor animal species was also reduced from 5 to 2 mg iodine/kg DM (EFSA 2013a; EFSA 2013b).

2.1.4.5. Milk Processing in the Dairy

Milk pasteurization (Norouzian et al. 2009) and ultra-high temperature processing (Stevenson, Drake & Givens 2018) are the most important steps of milk processing. Norouzian et al. (2009) and Norouzian (2011) used the high-temperature short time pasteurization-method (HTST) in two studies. In the first study, they measured an average decrease of iodine concentration in milk by 34% varying between 21.2 and 53.1%. The values for the second study were 27.4%, varying between 17.6 and 37.6%. Other authors also describe iodine losses during pasteurization of between 20 and 40% (Pedriali et al. 1997; Wheeler, Fleet & Ashley 1983), but Aumont et al. (1987) did not find any impact of pasteurization and spray drying on iodine concentration of milk.

Stevenson, Drake & Givens (2018) compared conventional milk with ultra-high temperature (UHT) processed milk from four supermarkets and found UHT milk to be 27% lower in iodine than conventional milk (427 vs. 314 µg/L; $P < 0.001$). Similar results are also reported by Arrizabalaga et al. (2015) and Payling et al. (2015).

2.1.5. Transfer of Iodine from Feed into Milk

The total excretion of iodine via milk (carry-over of iodine from feed into milk) varied between about 10% (with rape seed meal (Franke et al. 2009a) and up to 55% of iodine intake (Franke et al. 2009b; Swanson et al. 1990). On average, the iodine transfer from feed into milk amounted to 31.2% in the case of iodide and to 31.9% in the case of iodate (Franke 2009). In the Franke-studies (Franke 2009; Franke et al. 2009a) the carry-over of iodine decreased from 46.9% with diets without rapeseed meal, to 16.3% after supplementation with rapeseed.

3. EGGS

The eggs of laying hens may also contribute to about 10% of the total iodine intake of humans (see Section 2.4). Hester (2017) considered a high innovation potential of eggs for improving the I-supply in people. This section will deal with eggs as I-source for human nutrition and influencing factors on the I-content of eggs.

3.1. Iodine Content of "Standard-Eggs"

There is a large range in the iodine content presented in food composition tables and field analyses. The iodine content of whole hen eggs in the "Food Composition and Nutrition Tables" (Souci, Fachmann & Kraut 2008) was reported to be 85 – 100 µg/kg fresh weight (Table 4.3.1). The values for egg yolk vary between 75 – 160 µg while those for egg white are given with around 70 µg/kg fresh weight. Similar values for egg yolk (100 - 160 µg I/kg) and egg white (68 - 70 µg/kg) are reported in older references (e.g., Everson & Souders 1957).

Table 4.3.1: Iodine content in poultry eggs presented by various authors and food composition tables

Eggs or egg fractions	I-content (µg/kg fresh matter)	Authors
Whole eggs	93 (1 – 324)	CIEB (1952)
Whole egg	80 - 200 (World, 1964)	Vought, London & Brown (1964)
Whole egg	550 (UK, 1980)	Moxon & Dixon (1980)
Whole egg (Finland)	170	Varo et al. (1982)
Whole egg (UK)	525	Wenlock et al. (1982)
Whole egg	507	Mahesh, Deosthale & Rao (1992)
Whole egg	292	Anke, Groppel & Bauch (1993)
Whole egg	240	Dunn (1993)
Whole egg (USA)	480	Pennington et al. (1995)
Whole egg	505	Ministry of Agriculture Fisheries and Food (1997)
Whole egg	93	Fisher & Delange (1998)
Whole egg 1988 1994	 46[1] 600[2]	Anke et al. (2000)
Whole egg	480	Committee on toxicity of chemicals in food (COT) (2000a)
Whole egg	93 (World 1952)	Fordyce (2003)
Whole eggs	379 (4.4 - 4767[3])	Fordyce (2003)
Whole eggs (Switzerland)	324	Haldimann et al. (2005)
Egg yolk Egg white	1320 30	Hui & Sherkat (2006)
Whole egg Egg yolk Egg white	(85 - 98) 120 (75 - 160) 68	Souci, Fachmann & Kraut (2008)
Whole egg	630 (38 – 1000)	Rooke, Flockhart & Sparks (2010)
Whole egg Egg yolk Egg white	200 (77 - 1040) 1180 (520 - 3020) 54 (5 - 266)	Sager (2011)
Whole eggs (DM) 4 regions of Egypt	444 (261 – 518)	Hashish, Abdel-Samee & Abdel-Wahhab (2012)

[1] Without I-supplementation [2] up to 10 mg I/kg mixed feed
[3] With I-supplementation, based on data by (Kroupova et al. 1999)

Other authors give larger differences in the iodine content between egg yolk and egg white (e.g., Richter 1995; Hui & Sherkat 2006; Sager 2011; Röttger et al. 2012; Stibilj & Holcman 2002) (see Tables 4.3.1, 4.3.2, & 4.3.3). Differences between various authors and the high range in some references such as Fordyce (2003), Rooke, Flockhart & Sparks (2010) and Sager (2011) resulted from the feed of laying hens, for example feeds with goitrogenic substances, or differences in iodine supplementation of feeds.

3.2. Distribution of Iodine in Egg Yolk and Egg White (Albumen)

As already mentioned, there exist large differences in iodine deposition into egg yolk and egg albumen/white (see Tables 4.3.2 and 4.3.3). Cao et al. (1999) added extremely high amounts of iodine (50 and 100 mg/kg mixed feed) and found an increase of yolk iodine contents by 1395% and 1762%. In the albumin, iodine increased only by 121% and 187%. Lewis (2004) reviewed 13 studies with laying hens, where the authors did not add any iodine (basal content in the control group: 0.3 mg/kg) until up to 5 000 mg I/kg. The yolk iodine content varied between 6 and 7 000 µg/egg or between 100 and 117 000 µg/kg yolk.

Further dose-response studies (see Garber et al. 1993; Herzig & Suchy 1996; Kroupova et al. 1998; Kroupova et al. 1999; Rys et al. 1997; Saki et al. 2012) show significant influences of iodine supplementation on the iodine content of eggs. The objective of studies with very high I-supplementations (>20 mg I/kg) was to produce iodine enriched eggs, mainly in East Asia.

Richter (1995) supplemented up to 40 mg I per kg mixed feed and found more than 10 mg I per kg whole eggs (Table 4.3.2). The egg yolk contained nearly 20 mg I/kg. The I-recovery decreased from 18.6 with low supplementation to 11.6% in the highest supplemented group.

Yalcin et al. (2004) supplemented up to 24 mg I/kg mixed feed of laying hens and found Iodine concentrations in egg white similar to Richter (1995), but much lower values in egg yolk. They observed a linear increase in egg iodine from 160 to 1300 µg/kg whole egg in response to increases in dietary iodine supplementation of up to 21.5 mg/kg feed. Relevant data about the I-content in egg yolk, albumen and whole egg are shown in Table 4.3.3.

In consideration of various data mentioned above, the I-concentration in the yolk is between 2 (Souci, Fachmann & Kraut 2008) and 40 times higher than that of the albumen fraction. The average would consequently be about 10 times higher in the yolk (Röttger et al. 2012; Saki et al. 2012). Travnicek et al. (2006b) calculated a correlation coefficient between iodine contents of yolk and albumen of 0.67.

Table 4.3.2: Influence of various I-supplementations (KIO$_3$) for laying hens on the I-concentration in whole eggs, egg yolk and egg white as well as I-recovery in whole eggs (n = 3)

Iodine in feed (mg/kg)	Iodine in eggs (µg/kg fresh matter) Whole egg / Egg yolk / Egg white			I-recovery in eggs (% of I-intake)
0.4	140	350	17	18.6
0.9	330	950	28	17.3
5.4	1 460	5 750	140	12.5
20.4	7 000	18 500	600	16.3
40.4	10 670	19 250	1 260	11.6

Source: Richter (1995).

Table 4.3.3: Mean iodine concentrations of different samples (fresh matter) after Ca(IO$_3$)$_2$ supplementation (n = 6) on iodine concentration of yolk, albumin and whole egg

Iodine concentration of feed (mg/kg)	0.19	0.89	1.36	2.24	2.93	4.81
Egg yolk (µg/kg)	421	917	1258	1627	2244	3512
Egg albumen (µg/kg)	24	23	33	46	70	115
Whole egg (µg/kg)	131	296	415	525	711	1242

Source: Röttger et al. (2012).

3.3. Influencing Factors on the Iodine Content of Eggs

Apart from factors mentioned above, there are other influencing factors on the iodine content of eggs, such as level of iodine supplementation, duration of I-supplementation, iodine species, iodine antagonists in the feed, hen breed and management of hens. These are discussed next.

3.3.1. Level of I-supplementation of Feed

Laying hens excrete a considerable amount of ingested iodine within the eggs. This is the most important difference in iodine concentration in edible tissues of growing animals such as broilers, growing pigs, and ruminants on the one hand (Berk, Wagner & Flachowsky 2008; Franke et al. 2008; EFSA 2013a; EFSA 2013b; Herzig et al. 2005; Kaufmann & Rambeck 1998; Meyer et al. 2008; Röttger et al. 2011) (see Tables 4.4.1-4.4.4) and laying hens or lactating mammals on the other hand (e.g., EFSA 2013a; EFSA 2013b; Flachowsky et al. 2013; Flachowsky et al. 2017; Franke et al. 2009a; Franke et al. 2009b; Röttger et al. 2012; Schöne & Rajendram 2009) (see Tables 4.2.1-4.2.5). The data in Table 4.3.4 show a significant dose-response dependent relationship between iodine dosage and iodine content in the total eggs (24 eggs per group; studies with KI and $Ca(IO_3)_2$ and Lohmann White and Lohmann Brown hens).

Table 4.3.4: Iodine concentration of eggs (µg/kg complete eggs) depending on days of experimentation and level of iodine supplementation

Iodine Supplementation (mg/kg)	Duration of Supplementation (days)						
	0	4	8	15	29	85	164
0		119	112	122	100	131	196
0.25		158	185	166	147	237	350
0.50	140	127	195	225	240	294	457
2.5		209	638	600	657	742	964
5		380	1 146	1 347	1 363	1 433	1 600

Source: Flachowsky et al. (2017).

This significant influence of I-dose on the I-content of eggs has also been observed in other research (see Richter 1995; Rys et al. 1996a; Rys et al. 1996b; Kaufmann & Rambeck 1998; Lewis 2004; Yalcin et al. 2004; Röttger et al. 2012; Opalinski et al. 2012; Saki et al. 2012; Slupczynska et al. 2014 (as cited by Opalinski 2017).

Table 4.3.5 also indicates a dependence of I-concentration in eggs on the duration of feeding. At all dosage levels, there is an increase in iodine content of eggs until 15 days of experimentation, followed by a steady state for several weeks, then a further increase until 164 days. Long term studies (at least 150 days) are necessary to observe such effects. Slupczynska et al. (2014) measured 671 and 841 µg I/kg total eggs after three or five months of supplementation with 3 or 5 mg I/kg feed. The reason for this increase is not clear, but it could be associated with lower laying performance, higher feed intake of animals and higher proportion of egg yolk with longer laying period (Flachowsky et al. 2017). Similar results are reported by Szentirmai et al. (2013). The authors fractionated eggs of the lines TETRA brown and TETRA white layers at the age of 20 until 72 weeks. The egg weight increased from 48.3 (brown layers) and 42.4 (white layers) to >65 g for both genotypes, and the egg yolk ratio changed from 20.1/20.4% (brown/white) to >25% for both genotypes after 36 weeks of age until the end of the study. An increase of egg yolk ratio with duration of laying period was also observed in studies by Hartmann et al. (2000) with hens and by Applegate, Harper & Lilburn (1998) with ducks. Such changes may explain the increase in iodine content of whole eggs during experimentation period (see Table 4.3.4) and may influence the iodine content and the nutritive value of eggs in long term feeding studies.

3.3.2. Iodine Species

Researchers have studied differences in iodine content of eggs in relation to the iodine source (see Perry, Lewis & Hannagan 1989; Röttger et al. 2012). A higher iodine content with increasing iodine supplementation was found in groups that obtained KI compared with
$Ca(IO_3)_2$. Reasons for lower iodine content in eggs could include the lower intestinal uptake of iodate, since it has to be converted into iodine (Franke

2009; Lewis 2004; Moss & Miller 1970). Some authors report on a higher I-transfer from feed into eggs, if organic I-sources such as seaweed and kelp are included in the diet of the birds (Lewis 2004; Rys et al. 1996b).

Results reported in Table 4.3.6 demonstrate no clear effect of iodine source on iodine content of eggs especially after lower supplementations (5 mg/kg) effected higher I-concentration in eggs after supplementation of KI. The overall mean of the I-content of eggs was higher for KI (595 µg/kg) compared with $Ca(IO_3)_2$-supplementation (560 µg/kg total egg). Slupczynska et al. (2014) compared KI with KIO_3 (1, 3 and 5 mg/kg feed) without and with rapeseed meal in Hy-Line Brown Pullets over a period of 150 days. They found that the application of KI as I-source enhanced the accretion of iodine in eggs after 5 months of treatment (816 µg/kg) more efficiently than KIO_3 (698 µg/kg). Lewis (2004) reviewed the effects of high amounts of I-sources on the response of laying hens and did not find any clear relationship between organic and inorganic sources and between various inorganic sources such as KIO_3; $Ca(IO_3)_2$; KI and NaI.

3.3.3. Antagonists in Feed

Antagonists are substances, which may interact with nutrients and which can reduce the availability of certain nutrients. Rapeseed (Brassica napus) is the only important oilseed that can be grown on a commercial scale in Northern parts of the earth (e.g., in Canada, Northern Europe). Rapeseed and seeds or co-products of other cruciferea (e.g., canola, crambe, kale) contain valuable protein and amino acids, but they may also have deleterious effects on animal health and performance because of their goitrogen content (Schöne & Rajendram 2009). Goitrogens, such as glucosinolates are agents that may cause thyroid hypertrophy by interference with the thyroid hormone synthesis and secretion, but also with feedback mechanisms of thyroid-stimulating hormone and its releasing factor (Gaitan 1990; Tripathi & Mishra 2007).

Table 4.3.5: Iodine concentration of eggs (µg/kg complete eggs) depending on days of experimentation and level of iodine supplementation from KI and Ca(IO$_3$)$_2$

Iodine supplementation (mg I/kg from KI or Ca(IO$_3$)$_2$)	Duration of supplementation (days; KI/Ca(IO$_3$)$_2$ supplementation)						
	0	4	8	15	29	85	164
0[1]	121	119	112	121	100	131	196
0.25		179/137	202/169	177/136	155/166	254/220/	294/405
0.50		137/149	183/207	228/222	235/245	306/282	442/472
2.5		243/244	635/641	570/630	650/663	724/759	890/1038
5		447/384	1231/1060	1345/1263	1502/1224	1524/1242	1730/1470

Iodine content of basal diet: 0.35 mg/kg DM. Source: Flachowsky et al. (2017).

Table 4.3.6: Iodine concentration of eggs (µg/kg complete eggs) depending on days of experimentation, the level of iodine supplementation and diets without or with 10% rape seed cake- 24 eggs per group

Iodine supplementation (mg I/kg from KI or Ca(IO$_3$)$_2$)	Duration of supplementation (days; without/with 10% RSC) supplementation)						
	0	4	8	15	29	85	164
0[1]	121/136	119/129	112/102	122//81	100//86	131/111	196//124
0.25		158/127	185/141	167/135	160/131	237/181	350/196
0.50		145/122	195/174	225/170	240/163	294/195	457/228
2.5		243/203	638/539	600/532	657/536	742/555	964//648
5		415/287	1146/945	1347/962	1363/1042	1433/1113	1600//1197
Allover mean		216/174	455/380	492/376	504/392	567/431	713/479
+ RSC in % of control		80.6	83.5	82.6	77.8	76.0	67.1

[1] Iodine content of basal diet: 0.35 mg/kg DM Source: Flachowsky et al. (2017).

Table 4.3.7: Comparative analysis of iodine concentration of LSL and LB hen breeds

Iodine supplementation (mg I/kg from KI or Ca(IO$_3$)$_2$)	Duration of supplementation (days; LSL/LB hens)					
	4	8	15	29	85	164
0	104/135	106/117	105/138	81/116	112/150	172/220
0.25	129/187	166/205	158/176	153/168	207/267	319/381
0.50	127/160	176/214	208/243	206/274	271/317	412/502
2.5	210/277	547/728	576/625	581/732	645/838	839//1089
5	379/452	1026/1264	1266/1428	1286/1440	1310/1557	1395/1806
Allover mean	190/242	404/506	463/522	461/536	509//626	627/800
LSL in % of LB	78.5	79.8	88.7	86.0	81.3	78.4

Source: Flachowsky et al. (2017).

The negative effect of rapeseed and by-products from rapeseed (e.g., rapeseed cake; RSC) in iodine metabolism has been demonstrated in many studies as reviewed by Schöne & Rajendram (2009). Apart from the level of iodine supply, RSC also had a significant influence on the iodine content of eggs (see Table 4.3.6). RSC content in hen feed reduced the iodine concentration of eggs by an average of between 17% and 33%. Similar results are described by Goh & Clandinin (1977) and Slupczynska et al. (2014). Slupczynska et al. (2014) reported that after 3 to 5 months feeding, 10% of rapeseed meal reduced I-content in the whole egg from 671 to 841 µg/kg to 522 to 673 µg/kg. Lichovnikova & Zeman (2004) supplemented layer diets with 4.5, 9.0, and 13.5% rapeseed cake (containing 0.5; 0.99 and 1.49 µmol glucosinolates/g) and observed a positive effect on egg mass production and feed-conversion rate with higher levels of I-supplementation (3.6 compared with 6.1 mg I/kg feed).

Goitrogens from RSC did not only influence the iodine content in eggs (see Table 4.3.7), they also reduced the I-transfer of I from feed into eggs from an average of 23.9% without RSC to 15.6% with 10% RSC in hen diets (Table 4.3.9). Besides glucosinolates, the iodine metabolism in humans and animals is also influenced by the iodine dosage and the content of some trace elements like selenium, zinc, copper, iron and some ultra-trace elements like bromine, vanadium (Anke et al. 2000), as well as vitamin A deficiencies (Köhrle & Gärtner 2009) and steroid hormones, vitamin D and retinoid receptors (Davis et al. 2011; EFSA 2014).

3.3.4. Genetics: Hen Breeds

Measurements of the heritability of iodine and further trace nutrient concentrations in animal products are scarce (see Rooke, Flockhart & Sparks 2010). The influence of hen breed on iodine concentration in eggs has been observed in studies by Rys et al. (1997) and Röttger (2012). In a recent long-term study Flachowsky et al. (2017) showed clear influence of hen breed on iodine levels in eggs. In all phases of laying period and for all dosage levels, Lohmann Brown (LB) hens accumulated higher amounts of iodine in their eggs than Lohmann Selected Leghorn (LSL) hens.

Table 4.3.7 compares Iodine concentration of eggs (μg/kg complete eggs) of various breeds (LSL/LB-hens) depending on days of experimentation, the level of iodine supplementation from KI and Ca(IO$_3$)$_2$ and diets without RSC (12 eggs per group; Initial values: 124/131 μg/kg complete eggs for LSL/LB hens.

3.4. Animal Keeping and Farm Management

Stibilj & Holcman (2002) compared outdoor rearing with indoor keeping and found significant differential effects on iodine content in the white and the yolk fraction of eggs (Table 4.3.8). Indoor keeping resulted in higher I-concentration in the egg white and yolk fractions, probably because of the better I-supply.

Travnicek et al. (2006b) found a higher iodine content in eggs from large flocks (31.2 μg/egg corresponding to 500 μg/kg fresh weight) than in eggs from small flocks (10 μg/egg corresponding to 160 μg/kg fresh eggs). Similar proportions were seen in the iodine content of the yolk. The authors suggest that the differences may be caused by a higher iodine supplementation in commercial compound feed used on large farms.

Table 4.3.8: Iodine content in samples (μg/kg) of white and yolk fraction of eggs from free range and indoor keeping

Rearing of hens (mg I/kg feed)	Egg fraction	
	White	Yolk
Outdoor (0.18)	6.7 ± 2.2	114 ± 21
Indoor Study 1 (0.18)	7.5 ± 2.0	268 ± 42
Study 2 (0.66)	37.5 ± 3.9	740 ± 22

Source: Stibilj & Holcman (2002).

3.5. Transfer of Iodine from Feed into Eggs

The transfer values- that is, deposition in eggs as a percentage of iodine intake- of studies described above (Flachowsky et al. 2017; Richter 1995; Röttger et al. 2012) varies between 12.5% and 21.5% (Table 4.3.9). The transfer values were mainly influenced by the iodine supplementation- the lower the supplementation the higher the transfer values- and the portion of feeds containing antagonists, but also by the iodine source and the hen breed (Table 4.3.10).

Table 4.3.9: Carry over values of iodine from feed to eggs (% of added iodine in eggs)

Influencing factors	Transfer (% of iodine intake in eggs)
Iodine supplementation (mg/kg)	
0.25	21.5
0.5	18.1
2.5	12.8
5	12.5
Iodine source	
KI	18.0
$Ca(IO_3)_2$	16.0
Rape seed cake supplementation	
Without	19.6
+ 10% RSC	14.3
Hen breed	
Leghorn White (LSL)	15.3
Leghorn Brown (LB)	18.6

Source: Flachowsky et al. (2017).

3.6. Contribution of Eggs to I-Intake of Humans

The iodine content of eggs depends on many influencing factors as shown in the preceding discussion. Table 4.3.10 summarizes the influence of iodine content in feed as one of the most important factors for the iodine content in eggs, the relationship between egg consumption and the iodine intake from eggs, and the contribution to iodine requirements of humans. It shows the contribution of different numbers of daily consumed eggs (½, 1, 2, 3 eggs) to the daily iodine intake of consumers depending on iodine supplementation of the hen feed (1, 3, 5 mg I/kg feed), shown as total amount of iodine contributed by eggs and percentage shares of the daily iodine demand (150 µg/d for adults).

Table 4.3.10: Contribution of daily consumed eggs to the daily iodine intake of consumers depending on iodine supplementation of the hen feed

Iodine concentration of the hen feed [mg/kg]	1	3	5	1	3	5
Number of daily consumed eggs	Total amount of iodine contributed by eggs (µg/d)			% of daily demand (150 µg; see Table 4.2.1)		
½	7.4	18.6	29.8	4.9	12.4	19.9
1	14.8	37.2	59.6	9.9	24.8	39.7
2	29.5	74.3	119.1	19.7	49.5	79.4
3	44.3	111.5	178.7	29.5	74.3	119.1

Source: EFSA (2014).

1 mg I per kg mixed feed of hens and one egg per day may cover about 10% of the iodine requirements of adults (see Table 4.4.1). Travnicek et al. (2006b) concluded that one egg from large hen flocks cover 7 - 14% and from small flocks 2.2 - 4.4% of the daily iodine requirements of adults. Sager (2011) analysed eggs produced in Austria and found 31 µg I per egg, which is about 20% of iodine requirements of adults.

In consideration of results mentioned above, the FEEDAP-Panel of the European Food Safety Authority (EFSA 2013a; EFSA 2013b) proposed a reduction of the authorized maximum iodine concentration in complete feed for laying hens from 5 to 3 mg/kg feed with 88% DM. Under these conditions, eggs may be the second largest iodine source of animal origin for humans next to milk. The EU-commission (EU 2015) followed this proposal and recommended 3 mg I/kg feed as maximum content of total iodine in complete feed for laying hens, but still accepted exceptions up to a maximum content of 5 mg/kg. A reduction of iodine in mixed feed would not have any practical consequences in some European Union countries like Germany where field studies of iodine in mixed feed for laying hens varied between 0.5 and 2 mg I/kg (Grünewald, Steuer & Flachowsky 2006).

4. MEAT AND TISSUES OF LAND ANIMALS

Meat has a low iodine concentration (Tables 4.4.1-Table 4.4.4) and may contribute only small amounts to meeting human requirements. Most meat samples show I-values <100µg/kg fresh matter (Table 4.4.1).

In addition to results with broilers shown in Table 4.4.2, other studies also determined I-concentrations in poultry meat of <100 µg/kg, after <5 mg I/kg feed had been supplemented (Groppel, Rambeck & Gropp 1991; Hassanein, Anke & Hussein 2000; Kaufmann & Rambeck 1998; Röttger et al. 2012; Stibilj & Holcman 2002;). With the exception of geese (see Table 4.4.1), adequate values for meat of turkeys, ducks and other minor poultry species could not be found in the literature.

Similar results are also reported from experiments with growing pigs (Berk, Wagner & Flachowsky 2008; Franke et al. 2008; He, Hollwich & Rambeck 2002; Schöne et al. 2006b) (see Table 4.4.3) and field measurements in the Czech Republic (Herzig et al. 2005). Herzig et al. (2005) collected pork meat from 18 farms and found a range from 5 to 66 µg I/kg fresh matter with an average of 26 µg I/kg. Also, the "Food Composition and Nutrition Tables" by Souci, Fachmann & Kraut (2008) give similar data for pork (26-52 µg I/kg) (see Table 4.4.1).

Table 4.4.1: Iodine content of meat and meat products by various authors

Characterization	Origin	Mean (µg/kg)	Range (µg/kg)	Authors
Meat, beef/pigs	Chile	32	27 – 97	CIEB (1952)
Meat, meals	Chile		100 - 200	CIEB (1952)
Meat	Finland	<50		Varo et al. (1982)
Meat	UK	50	20 - 90	Wenlock et al. (1982)
Meat	USA	180	± 180	Pennington et al. (1995)
Meat	Switzerland, (2000 – 2001)	17	2 - 155	Haldimann et al. (2005)
Meat, poultry	Finland	90		Varo et al. (1982)
Meat	UK	75		Wenlock et al. (1982)
Meat	USA	170	± 190	Pennington et al. (1995)
Meat	Switzerland, 2000 – 2001	18	10 - 167	Haldimann et al. (2005)
Beef, muscle	Germany	54	17 - 68	Souci, Fachmann & Kraut (2008)
Bovine liver	Czech rep.	330 (in DM)		Mesko et al. (2010)
Beef, liver	Germany	140		Souci, Fachmann & Kraut (2008)
Mutton, leg	Germany	18		Souci, Fachmann & Kraut (2008)
Sheep, liver	Germany	33		Souci, Fachmann & Kraut (2008)
Veal, muscle	Germany	28		Souci, Fachmann & Kraut (2008)
Pork, muscle	Germany	45	26 - 52	Souci, Fachmann & Kraut (2008)
Pig`s, liver	Germany	140		Souci, Fachmann & Kraut (2008)
Goose, meat	Germany	40		Souci, Fachmann & Kraut (2008)

Table 4.4.2: Influence of I-supplementation to complete feed (KI and Ca(IO₃)₂) on I-content of body samples (µg/kg) of broilers

Iodine supplementation (mg/kg feed DM)	Control (0.5 mg/kg)	+ 1.0	+ 2.5	+ 5.0
Pectoral meat	5.8/7.0	10.3/11.5	39.3/27.9	58.0/52.1
Thigh meat	5.9/12.2	6.9/22.9	38.6/37.7	63.1/67.8
Liver	22.4/28.1	31.0/44.1	103.6/104.5	174.4/181.3

Source: Röttger et al. (2011).

Table 4.4.3: Influence of I-supplementation to complete feed on I-content of body samples of growing/fattening pigs

Iodine supplementation (mg/kg feed DM)	Control (0.5 mg/kg)	+ 1	+ 2	+ 5	+ 8
Meat (*Musc. long. dorsi*)	3.9	8,5	11	17	30
Liver	60	140	200	250	300

Source: Berk, Wagner & Flachowsky (2008); Franke et al. (2008).

Table 4.4.4: Influence of I-supplementation of feed (concentrate and maize silage) on I-content of body samples of growing/fattening bulls

Iodine concentration (mg/kg feed DM)	Control (0.79 mg/kg)	3.52	8.31
Musc. long. dorsi	16	45	80
Musc. glut. Medius	32	83	147
Liver	73	138	245
Kidney	93	231	450

Source: Meyer et al. (2008).

In a dose-response experiment with growing-fattening German Holstein bulls (223 – 550 kg body weight), Meyer et al. (2008) investigated the effect of various iodine doses (0.79; 3.52 and 8.31 mg I/ kg DM) on the iodine content of muscles *(M. longissimus dorsi; M. glutaeus medius)*, liver and kidney (Table 4.4.4). The iodine concentration in muscle, liver and kidney showed a statistically significant dose-related increase. However, compared with milk, the increase of iodine concentration in beef is very moderate (compare data in Table 4.2.2 with Table 4.4.4).

In agreement with data from Table 4.4.1, the liver also showed higher I-concentration than muscles in studies with broilers, pigs and bulls (see Tables 4.4.2 – 4.4.4). In contrast to milk (see Table 4.2.5) the data base shown here are not sufficient to compare the iodine content of meat and other slaughtered products after indoor and outdoor keeping or between organic and conventional production (Srednicka-Tober et al. 2016).

5. Fish

There exists a high variation in the iodine content of various fish species (see Table 4.5.1). Marine fish is the most important "classical" iodine source for human nutrition. The iodine content varies between 400 and 7 000 µg/kg fresh matter (see Table 4.5.1). Fresh-water fish contains much lower amounts of iodine than salt-water fish and is comparable to values of land animals (compare Tables 4.4.1-4.4.4 with Table 4.5.1). Depending on the portion of fish consumption, salt-water fish may contribute to a significant amount of human I-consumption.

Table 4.5.1: Iodine content of fish and fish products by various authors and food value tables

Characterisation; Fish species	Origin	Mean (µg/kg)	Range (µg/kg)	Authors
Fish meals	Chile		800 – 8 000	CIEB (1952)
Fish meals	Germany		1 450 – 3 320	Jeroch, Flachowsky & Weissbach (1993)
Fish, marine	Chile	832	163 – 1 440	CIEB (1952)
Fish, freshwater	Chile	30	17 – 40	CIEB (1952)
Fish, marine	Finland	460		Varo et al. (1982)
Fish, freshwater	Finland	165		Varo et al. (1982)

Characterisation; Fish species	Origin	Mean (µg/kg)	Range (µg/kg)	Authors
Fish, marine	UK	750	320 – 1 440	Wenlock et al. (1982)
Fish, marine	USA	1 160	(± 880)	Pennington et al. (1995)
Fish, marine	East Africa		1 117-5 310	Eckhoff & Maage (1997)
Fish, freshwater	East Africa		233 - 452	Eckhoff & Maage (1997)
Fish, marine	Switzerland, (2000 – 2001)	486	89 - 1593	Haldimann et al. (2005)
Fish, freshwater	Switzerland, (2000 – 2001)	98	3 - 408	Haldimann et al. (2005)
Fish, marine	Switzerland	2112	387 – 6 926	Haldimann et al. (2005)
Fish, freshwater	Switzerland	375	11 – 1 571	Haldimann et al. (2005)
Saltwater fish				
Flounder	Germany	260	44 – 1 540	Souci, Fachmann & Kraut (2008)
Halibut	Germany	370	220 - 520	Souci, Fachmann & Kraut (2008)
Atlantic Herring	Germany	470	240 - 670	Souci, Fachmann & Kraut (2008)
Cod	Germany	2 290	1 210 – 5 480	Souci, Fachmann & Kraut (2008)
Mackarel	Germany	500	390 - 820	Souci, Fachmann & Kraut (2008)
Mullet	Germany	3 300	1 600 – 4 900	Souci, Fachmann & Kraut (2008)
Redfish	Germay	350	300 – 1 240	Souci, Fachmann & Kraut (2008)
Sardine	Germany	320	130 - 540	Souci, Fachmann & Kraut (2008)
Haddock	Germany	1 350	600 – 5 100	Souci, Fachmann & Kraut (2008)

Table 4.5.1 (Continued)

Characterisation; Fish species	Origin	Mean (µg/kg)	Range (µg/kg)	Authors
Plaice	Germany	530	260 – 2 400	Souci, Fachmann & Kraut (2008)
Alaska Pollack	Germany	880	570 – 1 030	Souci, Fachmann & Kraut (2008)
Tuna	Germany	500	400 - 500	Souci, Fachmann & Kraut (2008)
Freshwater fish				
Eel	Germany	40		Souci, Fachmann & Kraut (2008)
Trout	Germany	35	30 - 36	Souci, Fachmann & Kraut (2008)
Carp	Germany	17		Souci, Fachmann & Kraut (2008)
Salmon	Germany	340		Souci, Fachmann & Kraut (2008)

CONCLUSION

There is considerable progress in the global iodine nutrition in humans and the "iodine status worldwide," although about 30 countries are still considered as iodine-deficient (<99 µg I/L urine). This progress has been achieved through programs of salt iodization, but also by iodine supplementation of animal feed. Milk, hen eggs and fish are very important iodine sources in human nutrition and are at the center of efforts to enhance iodine status of humans globally. The iodine content of foods, depends on many factors related to the diet of poultry and livestock, such as iodine content of feed, iodine source, iodine antagonists like glucosinolates in feed, farm management, among others. Meat, fresh water fish and other food of animal origin are less influenced by animal feeding and farm management but are poor in I-content.

There is no simple solution to declare the iodine concentration of milk, milk products and eggs on food labels. For this purpose, and taking into account the low range between requirements and upper levels (UL) for various human age groups (the range between requirements and UL for I is only 1: 2.5-3), iodine supplements should be used with caution in feed of lactating cows and other lactating ruminants and high iodine concentrations in milk should be avoided. Because of this, EFSA (EFSA 2013a; EFSA 2013b) proposed a reduction of the upper levels for iodine concentration in feed for dairy cattle and minor lactating ruminant species from 5 to 2 mg and in feed for laying hens from 5 to 3 mg iodine/kg feed.

Ongoing research in iodine sources and human intake is required to enhance our understanding of the use and misuse of this vital nutritional element. This is very important in the context of the overall global hunger problem and particularly the growing number of people worldwide who experience micro-nutrient deficiency.

REFERENCES

Anke, M, Glei, M, Rother, C, Vormann, J, Schäfer, U, Röhring, B, Drobner, C, Scholz, E, Hartmann, E, Möller, E & Sülze, A 2000, 'Die Versorgung Erwachsener Deutschlands mit Jod, Selen, Zink bzw. Vanadium und mögliche Interaktionen dieser Elemente mit dem Jodstoffwechsel,' [The supply of adult Germany with iodine, selenium, zinc or vanadium and possible interactions of these elements with the iodine metabolism] in K Bauch, (ed) *Aktuelle Aspekte des Jodmangels und Jodüberschusses. Interdisziplinäres Jodsymposium.* , pp. 147-176. Blackwell-Wiss. Verl., Berlin, Wien.

Anke, M, Groppel, B & Bauch, K-H 1993, 'Iodine in the Food Chain,' in F Delange, JT Dunn & D Glinoer, (eds), *Iodine Deficiency in Europe: A Continuing Concern*, pp. 151-158. Springer US, Boston, MA.

Applegate, TJ, Harper, D & Lilburn, MS 1998, 'Effect of hen production age on egg composition and embryo development in commercial Pekin ducks,' *Poultry Science*, vol. 77, no. 11, pp. 1608-1612. Available from: http://dx.doi.org/10.1093/ps/77.11.1608.

Arrizabalaga, JJ, Jalón, M, Espada, M, Cañas, M & Latorre, PM 2015, 'Concentración de yodo en la leche ultrapasteurizada de vaca. Aplicaciones en la práctica clínica y en la nutrición comunitaria,' ['Concentration of iodine in ultrapasteurized cow's milk. Applications in clinical practice and in community nutrition,] *Medicina Clínica*, vol. 145, no. 2, pp. 55-61. Available from: http://www.sciencedirect.com/science/article/pii/S0025775314005648.

Aumont, G, Le Querrec, F, Lamand, M & Tressol, JC 1987, 'Iodine Content of Dairy Milk in France in 1983 and 1984,' *Journal of Food Protection*, vol. 50, no. 6, pp. 490-493. Available from: http://jfoodprotection.org/doi/abs/10.4315/0362-028X-50.6.490.

Bath, SC, Button, S & Rayman, MP 2012, 'Iodine concentration of organic and conventional milk: implications for iodine intake,' *Br J Nutr*, vol. 107, no. 7, pp. 935-40.

Battaglia, M, Moschini, M, Giuberti, G, Gallo, A, Piva, G & Masoero, F 2009, 'Iodine carry over in dairy cows: effects of levels of diet fortification and milk yield,' *Italian Journal of Animal Science*, vol. 8, no. sup2, pp. 262-264. Available from: https://doi.org/10.4081/ijas.2009.s2.262.

Berg, JN & Padgitt, D 1985, 'Iodine Concentrations in Milk from Iodophor Teat Dips,' *Journal of Dairy Science*, vol. 68, no. 2, pp. 457-461. Available from: http://dx.doi.org/10.3168/jds.S0022-0302(85)80845-6.

Berk, A, Wagner, H & Flachowsky, G 2008, 'Influence of various amounts of glucosinolates in feed of pigs on the iodine content of thyroida and body samples of fattening pigs,' in *Forum für Angewandte Forschung*, Fulda, Germany, pp. 183-185.

Böhme, H, Kampf, D, Lebzien, P & Flachowsky, G 2005, 'Feeding value of crambe press cake and extracted meal as well as production responses of growing-finishing pigs and dairy cows fed these by-products,' *Arch Anim Nutr*, vol. 59, no. 2, pp. 111-22.

Bretthauer, EW, Mullen, AL & Moghissi, AA 1972, 'Milk Transfer Comparisons of Different Chemical Forms of Radioiodine,' *Health*

Physics, vol. 22, no. 3, pp. 257-260. Available from: http://journals. lww.com/health-physics/Fulltext/1972/03000/Milk_Transfer_ Comparisons_of_Different_Chemical.7.aspx.

Broadhead, GD, Pearson, IB & Wilson, GM 1965, 'Seasonal Changes in Iodine Metabolism. I. Iodine Content of Cows Milk,' *Bmj-British Medical Journal*, vol. 1, no. 5431, pp. 343-+. Available from: <Go to ISI>://WOS:A19656118700009.

Brzóska, F, Szybiński, Z & Śliwiński, B 2009, 'Iodine concentration in Polish milk - variations due to season and region,' *Endokrynologia Polska 60*, vol. 60, no. 6, pp. 449-454.

Cao, S, Yang, L, Cheng, M, Chen, L & Chen, K 1999, 'Effects of feeding high-iron and high-iodine diet to hens on the iron and iodide content in eggs and egg quality,' *Journal of Shanghai Agricultural College*, vol. 17, no. 4.

Castro, SI, Berthiaume, R, Robichaud, A & Lacasse, P 2012, 'Effects of iodine intake and teat-dipping practices on milk iodine concentrations in dairy cows,' *J Dairy Sci*, vol. 95, no. 1, pp. 213-20.

Castro, SI, Lacasse, P, Fouquet, A, Beraldin, F, Robichaud, A & Berthiaume, R 2011, 'Short communication: Feed iodine concentrations on farms with contrasting levels of iodine in milk,' *J Dairy Sci*, vol. 94, no. 9, pp. 4684-9.

Chacón Villanueva, C & Agency, CFS 2016, *Total diet study of iodine and the contribution of milk in the exposure of the catalan population, 2015*, MoH Government of Catalonia, Public Health Agency of Catalonia.

CIEB (Chilean Iodine Educational Bureau)1952, *Iodine content of foods; annotated bibliography, 1825-1951, with review and tables*, London.

Committee on toxicity of chemicals in food, 2000a, 'The 1997 Total Diet Study - Fluorine, Bromine, and Iodine (July 2000),' *Food Standards Agency*.

Committee on toxicity of chemicals in food, 2000b, 'Statement on iodine in cows' milk,' *Food Standards Agency*, pp. 1-5. Available from: https://cot.food.gov.uk/cotstatements/cotstatementsyrs/cotstatements20 00/cowsmilk.

Conrad, LM, III & Hemken, RW 1978, 'Milk Iodine as Influenced by an Iodophor Teat Dip,' *Journal of Dairy Science*, vol. 61, no. 6, pp. 776-780. Available from: http://dx.doi.org/10.3168/jds.S0022-0302(78)83648-0.

Crnkić, Ć, Haldimann, M, Hodzić, A & Tahirović, H 2015, *Seasonal and regional variations of the iodine content in milk from Federation of Bosnia and Herzegovina.*

Dahl, L, Johansson, L, Julshamn, K & Meltzer, HM 2004, 'The iodine content of Norwegian foods and diets,' *Public Health Nutrition*, vol. 7, no. 4, pp. 569-576. Available from: <Go to ISI>://WOS:000222034200014.

Dahl, L, Opsahl, JA, Meltzer, HM & Julshamn, K 2003, 'Iodine concentration in Norwegian milk and dairy products,' *British Journal of Nutrition*, vol. 90, no. 03, pp. 679-685. Available from: http://dx.doi.org/10.1017/S0007114503001740.

Davis, PJ, Lin, H-Y, Mousa, SA, Luidens, MK, Hercbergs, AA, Wehling, M & Davis, FB 2011, 'Overlapping nongenomic and genomic actions of thyroid hormone and steroids,' *Steroids*, vol. 76, no. 9, pp. 829-833. Available from: http://www.sciencedirect.com/science/article/pii/S0039128X11000523.

Diosady, LL, Alberti, JO, Venkatesh Mannar, MG & FitzGerald, S 1998, 'Stability of Iodine in Iodized Salt Used for Correction of Iodine-Deficiency Disorders. II,' *Food and Nutrition Bulletin*, vol. 19, no. 3, pp. 240-250. Available from: http://journals.sagepub.com/doi/abs/10.1177/156482659801900306.

Dunn, JT 1993, 'Sources of Dietary Iodine in Industrialized Countries,' in F Delange, JT Dunn & D Glinoer, (eds), *Iodine Deficiency in Europe*, pp. 17-23.

Eckhoff, K & Maage, A 1997, 'Iodine Content in Fish and Other Food Productsfrom East Africa Analyzed by ICP-M S,' *Journal of food composition and analysis 10,* pp. 270–282.

EFSA 2005, 'Opinion of the Scientific Panel on Additives and Products or Substances used in Animal Feed on the request from the Commission on the use of iodine in feedingstuffs,' *EFSA Journal*, vol. 168, no. 2, pp. 1-42.

EFSA 2013a, 'Scientific Opinion on the safety and efficacy of iodine compounds (E2) as feed additives for all animal species: calcium iodate anhydrous, based on a dossier submitted by Doxal Italia S.p.A,' *EFSA Journal*, vol. 11, no. 3, pp. 3178-n/a. Available from: http://dx.doi.org/10.2903/j.efsa.2013.3178.

EFSA 2013b, 'Scientific Opinion on the safety and efficacy of Iodine compounds (E2) as feed additives for all species: calcium iodate anhydrous (coated granulated preparation), based on a dossier submitted by HELM AG,' *EFSA Journal*, vol. 11, no. 2, pp. 3101-n/a. Available from: http://dx.doi.org/10.2903/j.efsa.2013.3101.

EFSA 2014, 'Scientific Opinion on Dietary Reference Values for iodine,' *EFSA Journal*, vol. 12(5), no. 3660, p. 57 pp.

EU 2015, 'Commission Implementing Regulation (EU) 2015/861 of 3 June 2015 concerning the authorisation of potassium iodide, calcium iodate anhydrous and coated granulated calcium iodate anhydrous as feed additives for all animal species (Text with EEA relevance),' *Official Journal of the European Union*, no. L 137/1.

Everson, GJ & Souders, HJ 1957, 'Composition and nutritive importance of eggs,' *J Am Diet Assoc*, vol. 33, no. 12, pp. 1244-54. Available from: http://www.ncbi.nlm.nih.gov/pubmed/13480800.

Falkenberg, U, Tenhagen, BA, Forderung, D & Heuwieser, W 2002, 'Effect of predipping with a iodophor teat disinfectant on iodine content of milk,' *Milchwissenschaft-Milk Science International*, vol. 57, no. 11-12, pp. 599-601. Available from: <Go to ISI>://WOS:000179842900001.

Fiedlerova, V 1998, 'Spectrophotometric determination of iodine and its content and stability in selected food raw materials and products,' *Czech Journal of Food Sciences*, no. 16, pp. 163-167.

Fisher, DA & Delange, FM 1998, 'Thyroid hormone and iodine requirements in man during brain development,' in JB Stannury, FM Delange, JT Dunn & CS Pandav, (eds), *Iodine in Pregnancy*. Oxford Univ. Publ., New Delhi.

Flachowsky, G, Franke, K, Meyer, U, Leiterer, M & Schöne, F 2013, 'Influencing factors on iodine content of cow milk.,' *European Journal of Nutrition*, pp. 1-15. Available from: http://dx.doi.org/10.1007/s00394-013-0597-4.

Flachowsky, G, Halle, I, Schultz, AS, Wagner, H & Dänicke, S 2017, 'Long term study on the effects of iodine sources and levels without and with rapeseed cake in the diet on the performance and the iodine transfer into body tissues and eggs of laying hens of two breeds,' *Landbauforschung Applied Agricultural Research*.

Flachowsky, G, Schone, F, Leiterer, M, Bemmann, D, Spolders, M & Lebzien, P 2007, 'Influence of an iodine depletion period and teat dipping on the iodine concentration in serum and milk of cows,' *Journal of Animal and Feed Sciences*, vol. 16, no. 1, pp. 18-25. Available from: <Go to ISI>://WOS:000244984700002.

Fordyce, FM 2003, 'Database of the Iodine Content of Food and Diets Populated with Data from Published Literature,' *British Geological Survey Commissioned Report, CR/03/84N*.

Franke, AA, Bruhn, JC & Osland, RB 1983, 'Factors Affecting Iodine Concentration of Milk of Individual Cows,' *Journal of Dairy Science*, vol. 66, no. 5, pp. 997-1002. Available from: http://dx.doi.org/10.3168/jds.S0022-0302(83)81894-3. [2017/12/06].

Franke, K 2009, *Effect of various iodine supplementations and species on the iodine transfer into milk and on serum, urinal and faecal iodine of dairy cows fed rations varying in the glucosinolate content Thesis (PhD)*, thesis, Martin-Luther-University, Halle-Wittenberg, Germany.

Franke, K, Meyer, U, Wagner, H & Flachowsky, G 2009a, 'Influence of various iodine supplementation levels and two different iodine species on the iodine content of the milk of cows fed rapeseed meal or distillers dried grains with solubles as the protein source,' *J Dairy Sci*, vol. 92, no. 9, pp. 4514-23.

Franke, K, Meyer, U, Wagner, H, Hoppen, HO & Flachowsky, G 2009b, 'Effect of various iodine supplementations, rapeseed meal application and two different iodine species on the iodine status and iodine excretion of dairy cows,' *Livestock Science*, vol. 125, no. 2, pp. 223-231.

Available from: http://www.sciencedirect.com/science/article/pii/S1871141309001747.

Franke, K, Schöne, F, Berk, A, Leiterer, M & Flachowsky, G 2008, 'Influence of dietary iodine on the iodine content of pork and the distribution of the trace element in the body,' *European journal of nutrition*, vol. 47, no. 1, pp. 40-46. Available from: https://www.openagrar.de/receive/fimport_mods_00000918 http://dx.doi.org/10.1007/s00394-007-0694-3.

French Agency for Food, (ANSES), 2005, 'Iodine.'

Gaitan, E 1990, 'Goitrogens in food and water,' *Annu Rev Nutr*, vol. 10, pp. 21-39.

Garber, DW, Henkin, Y, Osterlund, LC, Woolley, TW & Segrest, JP 1993, 'Thyroid function and other clinical chemistry parameters in subjects eating iodine-enriched eggs,' *Food and Chemical Toxicology*, vol. 31, no. 4, pp. 247-251. Available from: http://www.sciencedirect.com/science/article/pii/0278691593900749.

Goh, YK & Clandinin, DR 1977, 'Transfer of I125 to Eggs in Hens Fed on Diets Containing High-Glucosinolate and Low-Glucosinolate Rapeseed Meals,' *British Poultry Science*, vol. 18, no. 6, pp. 705-710. Available from: <Go to ISI>://WOS:A1977ED22100012.

Groppel, B, Rambeck, WA & Gropp, J 1991, 'Iodanreicherung in Organen und Geweben von Mastküken nach Iodsupplementation des Futters,' in *11. Arbeitstagung Mengen- und Spurenelemente*, Leipzig, pp. 300 - 308.

Grünewald, KH, Steuer, G & Flachowsky, G 2006, 'Field results of iodine in mixed feed,' in *9th Tagung Schweine- und Geflügelernährung*, Martin-Luther-Universität, Halle-Wittenberg, Halle, Germany, pp. 176-178.

Haldimann, M, Alt, A, Blanc, A & Blondeau, K 2005, 'Iodine content of food groups,' *Journal of Food Composition and Analysis*, vol. 18, no. 6, pp. 461-471. Available from: <Go to ISI>://WOS:000227850800001.

Hampel, R, Kairies, J & Below, H 2009, 'Beverage iodine levels in Germany,' *European Food Research and Technology*, vol. 229, no. 4, pp. 705-708. Available from: <Go to ISI>://WOS:000268275900019.

Hartmann, C, Johansson, K, Strandberg, E & Wilhelmson, M 2000, 'One-generation divergent selection on large and small yolk proportions in a White Leghorn line,' *British Poultry Science*, vol. 41, no. 3, pp. 280-286. Available from: <Go to ISI>://WOS:000088747100004.

Hashish, SM, Abdel-Samee, LD & Abdel-Wahhab, MA 2012, 'Mineral and Heavy Metals Content in Eggs of Local Hens at Different Geographic Areas in Egypt,' *Global Veterinaria*, vol. 8, no. 3, pp. 298-304. Available from: http://idosi.org/gv/GV8(3)12/15.pdf.

Hassanein, M, Anke, M & Hussein, L 2000, 'Determination of iodine content in traditional egyptian foods before and after a salt iodination programme,' *Polish journal of food and nutrition sciences*, vol. 9/50, no. 3, pp. 25-29.

Haug, A, Steinnes, E, Harstad, OM, Prestlokken, E, Schei, I & Salbu, B 2015, 'Trace elements in bovine milk from different regions in Norway,' *Acta Agriculturae Scandinavica Section a-Animal Science*, vol. 65, no. 2, pp. 85-96. Available from: <Go to ISI>://WOS: 000371034600003.

Haug, A, Taugbol, O, Prestlokken, E, Govasmark, E, Salbu, B, Schei, I, Harstad, OM & Wendel, C 2012, 'Iodine concentration in Norwegian milk has declined in the last decade,' *Acta Agriculturae Scandinavica Section a-Animal Science*, vol. 62, no. 3, pp. 127-134. Available from: <Go to ISI>://WOS:000317724700004.

He, ML, Hollwich, W & Rambeck, WA 2002, 'Supplementation of algae to the diet of pigs: a new possibility to improve the iodine content in the meat,' *Journal of Animal Physiology and Animal Nutrition*, vol. 86, no. 3-4, pp. 97-104. Available from: <Go to ISI>://WOS: 000174998500005.

Hejtmankova, A, Kuklik, L, Trnkova, E & Dragounova, H 2006, 'Iodine concentrations in cow's milk in Central and Northern Bohemia,' *Czech Journal of Animal Science*, vol. 51, no. 5, pp. 189-195. Available from: <Go to ISI>://WOS:000238002100002.

Hemken, RW, Vandersall, JH, Oskarsson, MA & Fryman, LR 1972, 'Iodine Intake Related to Milk Iodine and Performance of Dairy-Cattle,' *Journal of Dairy Science*, vol. 55, no. 7, pp. 931-+. Available from: <Go to ISI>://WOS:A1972M956100006.

Henderson, L & Gregory, J 2002, 'The National Diet & Nutrition Survey: adults aged 19 to 64 years,' *Office for National Statistics (ONS)*.

Herzig, I, Pisarikova, B, Kursa, J & Riha, J 1999, 'Defined iodine intake and changes of its concentration in urine and milk of dairy cows,' *Veterinarni Medicina*, vol. 44, no. 2, pp. 35-40. Available from: <Go to ISI>://WOS:000079768600002.

Herzig, I & Suchy, P 1996, 'Actual experience of importance iodine for animals,' *Veterinarni Medicina*, vol. 41, pp. 379-386.

Herzig, I, Travnicek, J, Kursa, J & Kroupova, V 2005, 'The content of iodine in pork,' *Veterinarni Medicina*, vol. 50, no. 12, pp. 521-525.

Hester, P 2017, *Egg Innovations and Strategies for Improvements*, 1st edn.

Hillman, D & Curtis, AR 1980, 'Chronic iodine toxicity in dairy cattle: blood chemistry, leukocytes, and milk iodide,' *J Dairy Sci*, vol. 63, no. 1, pp. 55-63.

Hui, YH & Sherkat, F 2006, *Handbook of Food Science, Technology, and Engineering*.

Jahreis, G, Leiterer, M & Fechner, A 2007, 'Jodmangelprophylaxe durch richtige Ernährung,' [Iodine deficiency prophylaxis through proper nutrition] *Prävention und Gesundheitsförderung*, vol. 2, no. 3, pp. 179-184. Available from: https://doi.org/10.1007/s11553-007-0068-y.

JECFA 1989, 'Toxicological evaluation of certain food additives and contaminants,' *WHO Food Additives Series*, vol. 24.

Jeroch, H, Flachowsky, G & Weissbach, F 1993, *Futtermittelkunde*, [Feed Science] Elsevier, München.

Johner, SA, Thamm, M, Nothlings, U & Remer, T 2013, 'Iodine status in preschool children and evaluation of major dietary iodine sources: a German experience,' *Eur J Nutr*, vol. 52, no. 7, pp. 1711-9. Available from: http://www.ncbi.nlm.nih.gov/pubmed/23212532.

Johner, SA, von Nida, K, Jahreis, G & Remer, T 2012, 'Time trends and seasonal variation of iodine content in German cow's milk - investigations from Northrhine-Westfalia,' *Berliner und Munchener*

Tierarztliche Wochenschrift, vol. 125, no. 1-2, pp. 76-82. Available from: <Go to ISI>://WOS:000299581400012.

Kaufmann, S & Rambeck, WA 1998, 'Iodine supplementation in chicken, pig and cow feed,' *Journal of Animal Physiology and Animal Nutrition-Zeitschrift Fur Tierphysiologie Tierernahrung Und Futtermittelkunde*, vol. 80, no. 2-5, pp. 147-152. Available from: <Go to ISI>://WOS:000078086300019.

Köhler, M, Fechner, A, Leiterer, M, Spörl, K, Remer, T, Schäfer, U & Jahreis, G 2012, 'Iodine content in milk from German cows and in human milk: new monitoring study,' *Trace Elements and Electrolytes*, vol. 29, no. 2, pp. 119-126.

Köhrle, J & Gärtner, R 2009, 'Selenium and thyroid,' *Best Pract Res Clin Endocrinol Metab*, vol. 23, no. 6, pp. 815-27.

Kroupova, V, Kratochvil, P, Kaufmann, S, Kursa, J & Travnicek, J 1998, 'Metabolic effect of iodine addition in laying hens,' *Veterinarni Medicina*, no. 43, pp. 207-212.

Kroupova, V, Travnicek, J, Kursa, J, Kratochvil, P & Krabacova, I 1999, 'Iodine content in egg yolk during its excessive intake by laying hens,' *Czech Journal of Animal Science*, vol. 44, no. 8, pp. 369-376. Available from: <Go to ISI>://WOS:000082352000006.

Kursa, J, Herzig, I, Trávníček, J & Kroupová, V 2005, 'Milk as a Food Source of Iodine for Human Consumption in the Czech Republic,' *Acta Veterinaria Brno*, vol. 74, pp. 255-264.

Lee, SM, Lewis, J, Buss, DH, Holcombe, GD & Lawrance, PR 1994, 'Iodine in British foods and diets,' *British Journal of Nutrition*, vol. 72, no. 3, pp. 435-446. Available from: <Go to ISI>://WOS: A1994PK29100009.

Lengemann, FW 1969, 'Radioiodine in the milk of cows and goats after oral administration of radioiodate and radioiodide,' *Health Phys*, vol. 17, no. 4, pp. 565-569.

Leskova, R 1969, 'Untersuchungen über die Jodversorgung der Milchrinder in Österreich,' ['Studies on the iodine supply of dairy cattle in Austria,'] *Wiener Tierärztliche Monatsschrift*, vol. 9, no. 56, pp. 369-74.

Lewis, PD 2004, 'Responses of domestic fowl to excess iodine: a review,' *British Journal of Nutrition*, vol. 91, no. 1, pp. 29-39. Available from: <Go to ISI>://WOS:000188966600005.

Lichovnikova, M & Zeman, L 2004, 'The effects of a higher amount of iodine supplement on the efficiency of laying hens fed extruded rapeseed and on eggshell quality,' *Czech Journal of Animal Science*, vol. 49, no. 5, pp. 199-203. Available from: <Go to ISI>://WOS:000222099400004.

Mahesh, DL, Deosthale, YG & Rao, BSN 1992, 'A sensitive kinetic assay for the determination of iodine in foodstuffs,' *Food Chemistry*, vol. 43, no. 1, pp. 51-56. Available from: http://www.sciencedirect.com/science/article/pii/030881469290241S.

Mesko, MF, Mello, PA, Bizzi, CA, Dressler, VL, Knapp, G & Flores, EM 2010, 'Iodine determination in food by inductively coupled plasma mass spectrometry after digestion by microwave-induced combustion,' *Anal Bioanal Chem*, vol. 398, no. 2, pp. 1125-31.

Meyer, U, Weigel, K, Schöne, F, Leiterer, M & Flachowsky, G 2008, 'Effect of dietary iodine on growth and iodine status of growing fattening bulls,' *Livestock Science*, vol. 115, no. 2-3, pp. 219-225. Available from: <Go to ISI>://WOS:000256168800013.

Miller, JK & Swanson, EW 1973, 'Metabolism of Ethylenediaminedihydriodide and Sodium or Potassium Iodide by Dairy Cows, *Journal of Dairy Science*, vol. 56, no. 3, pp. 378-384. Available from: http://dx.doi.org/10.3168/jds.S0022-0302(73)85181-1.

Ministry of Agriculture Fisheries and Food 1997, 'Dietary Intake of Iodine and Fatty Acids,' *Food Surveillance Information Sheet Number 127*.

Moschini, M, Battaglia, M, Beone, GM, Piva, G & Masoero, F 2009, 'Iodine and selenium carry over in milk and cheese in dairy cows: effect of diet supplementation and milk yield,' *animal*, vol. 4, no. 1, pp. 147-155. Available from: Cambridge Core. Available from: https://www.cambridge.org/core/article/iodine-and-selenium-carry-over-in-milk-and-cheese-in-dairy-cows-effect-of-diet-supplementation-and-milk-yield/AFE6303231515FFB08C949147B5EE66D.

Moss, BR & Miller, JK 1970, 'Metabolism of Sodium Iodide, Calcium Iodate, and Pentacalcium Orthoperiodate Initially Placed in Bovine Rumen or Abomasum,' *Journal of Dairy Science*, vol. 53, no. 6, pp. 772-&. Available from: <Go to ISI>://WOS:A1970G526300013.

Moxon, RED & Dixon, EJ 1980, 'Semi-automatic method for the determination of total iodine in food,' *Analyst*, vol. 105, no. 1249, pp. 344-352. Available from: http://dx.doi.org/10.1039/AN9800500344.

Norouzian, MA 2011, 'Iodine in raw and pasteurized milk of dairy cows fed different amounts of potassium iodide,' *Biol Trace Elem Res*, vol. 139, no. 2, pp. 160-7.

Norouzian, MA, Valizadeh, R, Azizi, F, Hedayati, M & Naserian, AA 2009, 'The Effect of Feeding Different Levels of Potassium Iodide on Performance, T-3 and T-4 Concentrations and Iodine Excretion in Holstein Dairy Cows,' *Journal of Animal and Veterinary Advances*, vol. 8, no. 1, pp. 111-114. Available from: <Go to ISI>://WOS: 000262562400023.

Nudda, A, Battacone, G, Decandia, M, Acciaro, M, Aghini-Lombardi, F, Frigeri, M & Pulina, G 2009, 'The effect of dietary iodine supplementation in dairy goats on milk production traits and milk iodine content,' *J Dairy Sci*, vol. 92, no. 10, pp. 5133-8.

Opalinski, S 2017, 'Supplemental iodin,' in PY Hester, (ed) *Egg innovations and strategies for improvements*, pp. 393-402.

Opalinski, S, Dolinska, B, Korczynski, M, Chojnacka, K, Dobrzanski, Z & Ryszka, F 2012, 'Effect of iodine-enriched yeast supplementation of diet on performance of laying hens, egg traits, and egg iodine content,' *Poultry Science*, vol. 91, no. 7, pp. 1627-1632. Available from: <Go to ISI>://WOS:000305590200017.

Paulikova, I, Seidel, H, Nagy, O & Kovac, G 2008, 'Milk Iodine Content in Slovakia,' *Acta Veterinaria Brno*, vol. 77, no. 4, pp. 533-538. Available from: <Go to ISI>://WOS:000263226200006.

Payling, LM, Juniper, DT, Drake, C, Rymer, C & Givens, DI 2015, 'Effect of milk type and processing on iodine concentration of organic and conventional winter milk at retail: Implications for nutrition,' *Food

Chemistry, vol. 178, no. Supplement C, pp. 327-330. Available from: http://www.sciencedirect.com/science/article/pii/S0308814615001053.

Pedriali, R, Giuliani, E, Margutti, A & Uberti, ED 1997, 'Iodine assay in cow milk - Industrial treatments and iodine concentration.,' *Annali Di Chimica*, vol. 87, no. 7-8, pp. 449-456. Available from: <Go to ISI>://WOS:A1997XV46600002.

Pennington, JAT, Schoen, SA, Salmon, GD, Young, B, Johnson, RD & Marts, RW 1995, 'Composition of core foods of the US food supply, 1982-1991,' *Journal of Food Composition Analysis*, no. 8, pp. 171-217.

Perry, G, Lewis, PD & Hannagan, MJ 1989, 'Iodine supplementation from two sources and its effect on egg output,' *British Poultry Science* no. 30, pp. 973-974.

Ranz, D, Scherer-Herr, K & Rambeck, WA 2001, 'Iodine secretions in urine and milk in a dairy farm in Nordrhein-Westfalia,' *Proceedings - society of nutrition physiology*, p. 67.

Rasmussen, LB, Larsen, EH & Ovesen, L 2000, 'Iodine content in drinking water and other beverages in Denmark,' *European Journal of Clinical Nutrition*, vol. 54, no. 1, pp. 57-60. Available from: <Go to ISI>://WOS:000085395400011.

Rasmussen, MD, Galton, DM & Petersson, LG 1991, 'Effects of Premilking Teat Preparation on Spores of Anaerobes, Bacteria, and Iodine Residues in Milk,' *Journal of Dairy Science*, vol. 74, no. 8, pp. 2472-2478. Available from: <Go to ISI>://WOS:A1991FZ99800013.

Rey-Crespo, F, Miranda, M & Lopez-Alonso, M 2013, 'Essential trace and toxic element concentrations in organic and conventional milk in NW Spain,' *Food and Chemical Toxicology*, vol. 55, pp. 513-518. Available from: <Go to ISI>://WOS:000317536900064.

Richter, G 1995, 'Einfluss der Jodversorgung auf den Jodgehalt im Ei,' [Influence of iodine supply on the iodine content in the egg] in *15. Arbeitstagung, Mengen- & Spurenelemente*, Friedrich-Schiller-Univerität, Jena, pp. 457-64.

Rooke, JA, Flockhart, JF & Sparks, NH 2010, 'The potential for increasing the concentrations of micro-nutrients relevant to human nutrition in meat, milk and eggs,' *Journal of Agricultural Science*, vol. 148, pp. 603-614. Available from: <Go to ISI>://WOS:000282621500010.

Röttger, AS 2012, *The effect of various iodine sources and levels on the performance and the iodine transfer in poultry products and tissues*, thesis, Tierärztliche Hochschule Hannover, Germany.

Röttger, AS, Halle, I, Wagner, H, Breves, G, Daenicke, S & Flachowsky, G 2012, 'The effects of iodine level and source on iodine carry-over in eggs and body tissues of laying hens,' *Archives of Animal Nutrition*, vol. 66, no. 5, pp. 385-401. Available from: <Go to ISI>://WOS: 000308726000004.

Röttger, AS, Halle, I, Wagner, H, Breves, G & Flachowsky, G 2011, 'The effect of various iodine supplementations and two different iodine sources on performance and iodine concentrations in different tissues of broilers,' *British poultry science*, vol. 52, no. 1, pp. 115-123. Available from: https://www.openagrar.de/receive/import_mods_00000412 http://www.tandfonline.com/doi/abs/10.1080/00071668.2010.539591?url_ver=Z39.88-2003&rfr_id=ori:rid:crossref.org&rfr_dat=cr_pub%3dpubmed#.U0KlWfmqmfU http://dx.doi.org/10.1080/00071668.2010.539591.

Rozenska, L, Hejtmankova, A, Kolihova, D & Miholova, D 2011, 'Selenium and iodine content in sheep milk from farms in central and east Bohemia,' *Scientia Agriculturae Bohemica*, no. 42, pp. 153-158.

Rys, R, Weir-Konas, E, Pyska, H & Kuchta, M 1996a, 'The effect of different types and levels of iodine additives in feeds on iodine deposition in eggs,' *Roczniki Naukowe Zootechniki* vol. 23, no. 2, pp. 187-197.

Rys, R, Wir-Konas, E, Pyska, H, Kuchta, M & Pietras, M 1996b, 'Effect of varying amounts of kelp and KI in the diets on the laying performance, iodine and cholesterol content of eggs, and thyroid hormones in the blood of layers,' *Roczniki Naukowe Zootechniki* vol. 23, no. 2, pp. 199-214.

Rys, R, Wir-Konas, E, Pyska, H, Kuchta, M & Pietras, M 1997, 'Changes in egg iodine concentration in three hen strains in relation to iodine level in diets,' *Roczniki Naukowe Zootechnik* vol. 24, pp. 229-42.

Rysava, L, Kubackova, J & Stransky, M 2007, 'Jod- und Selengehalte in der Milch aus neun europäischen Ländern,' ['Iodine and selenium in milk from nine European countries,'] *Proceedings German Nutrition Society (DGE); Bonn, Germany*, pp. 45-48.

Sager, M 2011, *Major and trace elements in hens' eggs from Austria*.

Saki, AA, Aslani Farisar, M, Aliarabi, H, Zamani, P & Abbasinezhad, M 2012, *Iodine enriched egg production in response to dietary iodine in laying hens*.

Schöne, F, Leiterer, M, Lebzien, P, Bemmann, D, Spolders, M & Flachowsky, G 2009, 'Iodine concentration of milk in a dose-response study with dairy cows and implications for consumer iodine intake,' *J Trace Elem Med Biol*, vol. 23, no. 2, pp. 84-92.

Schöne, F & Rajendram, R 2009, 'Iodine in farm animals,' in VR Preedy, GN Burrow & RR Watson, (eds), *Comprehensive Handbook of Iodine. Nutritional, Pathological and Therapeutic Aspects*, pp. 151-70 Oxford.

Schöne, F, Sperrhake, K, Engelhard, T & Leiterer, M 2006a, 'Jodkonzentration der Milch unter dem Einfluss von Rapsextraktionsschrot im Futter,' ['Iodine concentration of milk under the influence of rapeseed meal in feed,'] *118. VDLUFA-Kongress in Freiburg, Germany*, p. 44.

Schöne, F, Sporl, K & Leiterer, M 2017, 'Iodine in the feed of cows and in the milk with a view to the consumer's iodine supply,' *Journal of Trace Elements in Medicine and Biology*, vol. 39, pp. 202-209. Available from: <Go to ISI>://WOS:000390505900029.

Schöne, F, Zimmermann, C, Quanz, G, Richter, G & Leiterer, M 2006b, 'A high dietary iodine increases thyroid iodine stores and iodine concentration in blood serum but has little effect on muscle iodine content in pigs,' *Meat Science*, vol. 72, no. 2, pp. 365-372. Available from: http://www.sciencedirect.com/science/article/pii/S030917400500327X.

Slupczynska, M, Jamroz, D, Orda, J & Wiliczkiewicz, A 2014, 'Effect of various sources and levels of iodine, as well as the kind of diet, on the performance of young laying hens, iodine accumulation in eggs, egg characteristics, and morphotic and biochemical indices in blood,'

Poultry Science, vol. 93, no. 10, pp. 2536-2547. Available from: <Go to ISI>://WOS:000343697800013.

Soriguer, F, Gutierrez-Repiso, C, Gonzalez-Romero, S, Olveira, G, Garriga, MJ, Velasco, I, Santiago, P, de Escobar, GM & Garcia-Fuentes, E 2011, 'Iodine concentration in cow's milk and its relation with urinary iodine concentrations in the population,' Clinical Nutrition, vol. 30, no. 1, pp. 44-48. Available from: http://www.sciencedirect.com/science/article/pii/S0261561410001305.

Souci, SW, Fachmann, W & Kraut, H 2008, Food Composition and Nutrition Tables: Die Zusammensetzung der Lebensmittel, Nährwert-Tabellen La composition des aliments Tableaux des valeurs nutritives, 7 ed., Wissenschaftliche Verlagsgesellschaft.

Srednicka-Tober, D, Baranski, M, Seal, C, Sanderson, R, Benbrook, C, Steinshamn, H, Gromadzka-Ostrowska, J, Rembialkowska, E, Skwarlo-Sonta, K, Eyre, M, Cozzi, G, Larsen, MK, Jordon, T, Niggli, U, Sakowski, T, Calder, PC, Burdge, GC, Sotiraki, S, Stefanakis, A, Yolcu, H, Stergiadis, S, Chatzidimitriou, E, Butler, G, Stewart, G & Leifert, C 2016, 'Composition differences between organic and conventional meat: a systematic literature review and meta-analysis,' British Journal of Nutrition, vol. 115, no. 6, pp. 994-1011. Available from: <Go to ISI>://WOS:000372155800007.

Stevenson, MC, Drake, C & Givens, DI 2018, 'Further studies on the iodine concentration of conventional, organic and UHT semi-skimmed milk at retail in the UK,' Food Chem, vol. 239, pp. 551-555.

Stibilj, V & Holcman, A 2002, 'Se and I content of eggs and tissues of hens from farmyard rearing,' 21. Arbeitstagung Mengen und Spurenelemente, Jena, pp. 437-443.

Swanson, EW, Miller, JK, Mueller, FJ, Patton, CS, Bacon, JA & Ramsey, N 1990, 'Iodine in milk and meat of dairy cows fed different amounts of potassium iodide or ethylenediamine dihydroiodide,' J Dairy Sci, vol. 73, no. 2, pp. 398-405.

Szentirmai, E, Milisits, G, Donko, T, Budai, Z, Ujvari, J, Fulop, T, Repa, I & Suto, Z 2013, 'Comparison of changes in production and egg composition in relation to in vivo estimates of body weight and composition of brown and white egg layers during the first egg-laying period,' *British Poultry Science*, vol. 54, no. 5, pp. 587-593. Available from: <Go to ISI>://WOS:000327513700006.

Travnicek, J, Herzig, I, Kursa, J, Kroupova, V & Navratilova, M 2006a, 'Iodine content in raw milk,' *Veterinarni Medicina*, vol. 51, no. 9, pp. 448-453. Available from: <Go to ISI>://WOS:000240880200002.

Travnicek, J, Kroupova, V, Herzig, I & Kursa, J 2006b, 'Iodine content in consumer hen eggs,' *Veterinarni Medicina*, vol. 51, no. 3, pp. 93-100. Available from: <Go to ISI>://WOS:000236973000003.

Travnicek, J & Kursa, J 2001, 'Iodine concentration in milk of sheep and goats from farms in south Bohemia,' *Acta Veterinaria Brno*, vol. 70, no. 1, pp. 35-+. Available from: <Go to ISI>://WOS:000168076000006.

Tripathi, MK & Mishra, AS 2007, 'Glucosinolates in animal nutrition: A review,' *Animal Feed Science and Technology*, vol. 132, no. 1-2, pp. 1-27. Available from: <Go to ISI>://WOS:000242776300001.

Troan, G, Dahl, L, Meltzer, HM, Abel, MH, Indahl, UG, Haug, A & Prestlokken, E 2015, 'A model to secure a stable iodine concentration in milk,' *Food & Nutrition Research*, vol. 59. Available from: <Go to ISI>://WOS:000367154300001.

Varo, P, Saari, E, Paaso, A & Koivistoinen, P 1982, 'Iodine in Finnish Foods,' *International Journal for Vitamin and Nutrition Research*, vol. 52, no. 1, pp. 80-89. Available from: <Go to ISI>://WOS:A1982NM92100013.

Vought, RL, London, WT & Brown, FA 1964, 'A Note on Atmospheric Iodine and Its Absorption in Man,' *J Clin Endocrinol Metab*, vol. 24, pp. 414-6. Available from: http://www.ncbi.nlm.nih.gov/pubmed/14169498.

Waszkowiak, K & Szymandera-Buszka, K 2008, 'Effect of storage conditions on potassium iodide stability in iodised table salt and collagen preparations,' *International Journal of Food Science and Technology*, vol. 43, no. 5, pp. 895-899. Available from: <Go to ISI>://WOS:000254808800019.

Wenlock, RW, Buss, DH, Moxon, RE & Bunton, NG 1982, 'Trace nutrients. 4. Iodine in British food,' *Br J Nutr*, vol. 47, no. 3, pp. 381-90. Available from: http://www.ncbi.nlm.nih.gov/pubmed/7082612.

Wheeler, SM, Fleet, GH & Ashley, RJ 1983, 'Effect of Processing Upon Concentration and Distribution of Natural and Iodophor-Derived Iodine in Milk,' *Journal of Dairy Science*, vol. 66, no. 2, pp. 187-195. Available from: <Go to ISI>://WOS:A1983QD39900001.

Yalcin, S, Kahraman, Z, Yalcin, S, Yalcin, SS & Dedeoglu, HE 2004, 'Effects of supplementary iodine on the performance and egg traits of laying hens,' *British Poultry Science*, vol. 45, no. 4, pp. 499-503. Available from: <Go to ISI>://WOS:000224024600008.

In: Agriculture, Food, and Food Security
Editor: Clinton Lloyd Beckford
ISBN: 978-1-53613-483-4
© 2018 Nova Science Publishers, Inc.

Chapter 5

UNLOCKING THE FULL POTENTIAL OF CARAMBOLA (*AVERRHOA CARAMBOLA*) AS A FOOD SOURCE: BOTANY, GROWING, PHYSIOLOGY AND POSTHARVEST TECHNOLOGY

Noureddine Benkeblia[*]
Department of Life Sciences/Biotehchonoly Centre,
University of the West Indies, Mona Campus, Kingston, Jamaica

1. BACKGROUND

Tropical crops, including fruits, play an important role in the daily diets of billions of people around the world. Many tropical fruits are harvested from wild, or locally cultivated trees of a wide range of minor and under-utilized species. Carambola is well known in the United States, especially in California and Florida, and Europe, and is also widely consumed in the

---

[*] Corresponding author email: noureddine.benkeblia@uwimona.edu.jm.

Caribbean and Asia, where it originated. Unfortunately, carambola remains underutilized because like many other tropical fruits, it has undergone very little scientific improvements. Moreover, carambola, like many other underutilized fruits which are well adapted to the local climates, are highly nutritious and might contribute significantly to enhance diet and food and nutrition security of rural people and play a significant role in nutritional status and socio-economic development of rural communities in many countries especially in the tropics. This chapter discusses issues relevant to the production, utilisation and postharvest handling of carambola in order to highlight the scientific status on this fruit and to summarize the nutritional and technological data which is necessary to capitalize on the considerable potential of carambola.

2. INTRODUCTION

The fruit of carambola (*Averrhoa carambola*) belongs to the Oxalidaceae, and has an interesting history as the fruit diffused around the world acquiring many regional names in the process. Carambola, also known as Starfruit, because of its distinct star-shape when cut, is a tropical fruit, with a sweet-sour taste. It is a popular delicacy in many countries and is most often eaten as fresh fruit.

The carambola tree is thought to be native to Asia, and has been cultivated in Southeast Asia and Malaysia for over three hundred years (Morton 1987). However, due to the health properties of its fruit, the tree diffused rapidly to many other countries especially in the tropics and sub-tropics. The tree has been popular in certain parts of the United States with tropical climate– mainly in Hawaii and Florida – for about a hundred years, and it grows across the Caribbean and some South American countries as well. Carambola has a growing season – from blossoming to fruit harvesting-that extends from July to February in the Caribbean and Latin America.

In addition to its popular Spanish name 'carambola' it is also known as *balimbing, belimbing*, or *belimbing manis* ("sweet belimbing"), *kamaranga*, *kamruk* in the Orient likely China. It is also called *kamrakhin* in Sri-Lanka and India, *Khe* or *khe tain* in Vietnam, *nak fuang* in Laos, *carambolier* in France, *ma fueang* in Thailand, *belimbing batu, belimbing besi, belimbing pessegi, belimbing sayur, belimbing saji, kambola, caramba,* or *star fruit* by Malayans. In the Caribbean, carambola is also known by different names including *five fingers* in Guyana, *vinagrillo* in the Dominican Republic, *zibline* in Haiti, *cornichon* in the French Antilles, while in Trinidad, it is called *coolie tamarind* (Manda et al. 2012).

3. ORIGIN AND DISTRIBUTION

Although extensive literature exists on carambola fruit, its exact origin is still a matter of disagreement. One school of thought suggests that it may have originated in Sri-Lanka and the Moluccas (Chandler 1958) although it has been cultivated in Southeast Asia and Malaysia for many centuries. Another view holds that it could be native to Indonesia where a wild species has been found (Purseglove 1968) and/or southern China where it has been cultivated for many centuries, and it is commonly grown in the provinces of Fukien, Kuangtung and Kuangsi. Some literature suggests it is native to the Malayan archipelago where the two species *Averrhoa carambola* and *Averrhoa bilimbi* have been found (Burkill 1966; Popenoe 1920). Carambola fruit has never been located in the wild except in Indonesia (Purseglove 1968) and according to some sources, it was thought to have been domesticated throughout India and Southeast Asia in prehistoric times. Since 1800s, the tree has diffused widely to become very popular in the Philippines and Australia, and moderately popular in some South Pacific Islands, particularly Tahiti, New Caledonia and Netherlands New Guinea, and in Guam, while it is also well known in Taiwan, India, South America, Florida, Hawaii, and the Caribbean region.

The carambola trees are cultivated throughout many tropical and warm subtropical regions, and some specimens of the tree are found in special collections in the Caribbean islands, Central America, tropical South America, and also in parts of West Africa and Zanzibar. Several trees were grown from 1935 in Israel, but in many areas, trees are grown more as an ornamental than fruit tree. By the end of the 1800s, the plant was introduced into southern Florida and was viewed mainly as a curiosity until recent years when its fruit was used for decoration (Campbell et al. 1985).

4. Description

The carambola tree is considered as a slow-growing tree of short single or multi-trunk with a much-branched, bushy, broad, rounded crown and can grow up to nine meters in height, and spreads six to eight meters wide. The leaves are deciduous, spirally arranged, and alternate, imparipinnate, 15-20 cm long, with 5 to 11 nearly opposite leaflets, ovate or ovate-oblong, four to nine cm long; soft, medium-green, and smooth on the upper surface, finely hairy and whitish on the underside (Campbell 1965; Ruehle, 1958). The plant, which flowers and fruits all year round in some regions like the Caribbean, have small clusters of red-stalked, lilac, purple-streaked, downy flowers, about six mm wide which are borne on the twigs in the leaves' axils, however, fruit borne on very young twigs is considered of poor quality compared to those borne on older twigs (Knight 1965; Knight 1982). However, the flowering of carambola trees seems to be more influenced by the cultivars and water stress rather than by temperatures and photoperiod (Salakpetch et al. 1990).

The showy, oblong, longitudinally five to six angled fruits, six to 15 cm long and c.a. eight to nine cm wide, have thin, waxy, orange-yellow skin and juicy, crisp, yellow flesh when fully ripe. The fruit have a more or less pronounced oxalic acid odour and the flavour ranges from very sour to very sweet, and slices cut in cross-section have the form of a star. Good carambola cultivars have an agreeable, acid to sweet flavour depending on the cultivars and ripening stages. Fruit are sweetest when allowed to ripen on the tree. Green fruit will slowly turn yellow if picked before fully ripe, and it takes about 60 to 75 days from fruit set to maturity depending upon cultivar, production practices, and weather (Campbell 1965).

The fruit is ovoid to ellipsoid, with 4-6 prominent longitudinal ribs corresponding to the carpels. Length of the fruit is 4-6 inches or more, with colour varying from whitish to a deep golden yellow. The skin is smooth, translucent and waxy. The pulp is juicy, varying in texture from very soft to firm and crisp. Flavour varies from very sour with little sugar to very sweet with little acidity (Berry 1977; Shui & Leong 2004a, 2004b).

The carambola fruit may contain between one to 23 seeds but usually contains four to five seeds although seedless cultivars are found. Seeds are 0.6 to 1.3 cm long, thin, light brown, and enclosed by a gelatinous aril (Morton 1987).

Figure 5.1. Carambola (Averrhoa carambola) tree (A), leaves and flowers (B), and ripe fruits (C).

5. CULTIVARS

There are two distinct classes of carambola: the smaller very sour type, flavoured and containing more oxalic acid, and the larger, called "sweet" type which is less mild-flavoured, rather bland and containing less oxalic acid. The most known sweet varieties are 'Arkin', 'Fwang Tung', 'Kari' and 'Sri Kembangan'.

In 1935, seeds from Hawaii were planted in Florida and a selection from the resulting seedlings was vegetatively propagated during the 1940's and 1950's. In the late 1960s, a recognized variety was released under the name 'Golden Star' and distributed to growers in Florida. This variety has a large, deeply winged, decorative, and mildly sub-acid to sweet. Presently, there are many cultivars in USA, mainly in Florida, however, few are available in nurseries and many still have not been evaluated thoroughly for commercial production potential. The most important features of commercial cultivars include high productivity, good eating quality, medium fruit size, good yellow colour, resistance to damage, and ability to maintain good quality during storage and marketing (Wagner et al. 1975). The principal cultivars that are in commercial production include 'Arkin', 'Fwang Tung', 'Golden Star', 'Maha', 'Newcomb', and 'Thayer'. Presently, 'Arkin' is the most popular cultivar because of its excellent flavour and resistance to handling damage. The 'Golden Star', 'Newcomb', and 'Thayer' are widely planted also because they are productive, have good colour, and are easy to market, while 'Twang Tung' and 'Maha' are not popular because their fruit has poor colour, is easily damaged in handling, and seems to be susceptible to damage in refrigerated storage (Campbell et al. 1985). Several cultivars are grown in other countries, for example 'Mih Tao', 'Dah Pon' and 'Tean Ma' in Taiwan, and the varieties 'Fwang Tung' in Thailand.

6. GROWING

The carambola tree is well-suited to the sub-tropical and tropical climates from sea level to 600 meters. Trees need high rainfall ranging between 1500 and 3000 mm annually. When the dry season extends more than two months, it will impact negatively on plant growth and fruit production. Trees grow very well in friable and well-drained soil of pH ranging 5.5 to 6.5. The trees can also grow well in other various types of soil from sandy to clayey, however, they need a good soil improvement and management regime especially irrigation and application of fertiliser. Carambola trees do not tolerate conditions such as drought, flooding (Ismail & Noor 1996a) and high salinity. All carambola cultivars do not require complete shade to grow but, shade up to 40-50% is considered ideal.

Trees grown from seeds may not flower until they are four to six years old, while grafted trees begin to flower at nine months. Depending upon variety, fruit ripens 80-110 days after anthesis but, harvesting may begin about 65 days after anthesis where fruit set is considered good. Later, fruit fall tends to limit the crop. Ismail and Noor (1996b) reported that root growth restriction and appropriate irrigation was beneficial to carambola trees growing and hastening flower initiation. Moreover, de-fruiting and selective pruning of trees increase early-season fruit production (Núñez-Elisa & Crane 2000).

In orchards, carambola trees' susceptibility to wind damage varies based upon cultivars and windbreaks are often recommended on exposed sites. When damaged by winds, trees show different symptoms including defoliation, desiccation, twig dieback, stunted growth, and wind scar (fruit damage). During flowering, heavy rainfall may affect pollination and fruit production. Trees are also susceptible to root rot under wet conditions, but generally do well with moderate, year-round rainfall. Trees planted in humid environments are susceptible to algae and the fruit is prone to anthracnose fruit spot (Campbell 1965; Crane 1994).

7. COMPOSITION AND NUTRITIVE VALUE

The nutritional composition of carambola fruit varies with maturity and ripening stages, and quite a difference was found between mature green and mature yellow fruits. Carambola fruit is low in calories (c.a. 36 to 57 cal $100g^{-1}$), and considered as a good source of potassium, and a moderate to good source of vitamin C (Berry 1977; Wenkam & Miller 1965). The total solids of carambola vary from five to 14%, and the acidity also varies from two to ten mg 100 g^{-1} tissue. Other flavour components make considerable variation in the palatability of fruit of different cultivars (Berry 1977).

The composition of the fruit was determined as shown by Table 5.1. As reported by many authors, the chemical and physical characteristics of carambola fruit measured over the harvest season, vary with the ripening stages, and it was noted that total soluble solids, total sugars, and total ascorbic acid content increased whereas total titrable acidity declined (Ali & Jaafar 1992; Patil et al. 2010).

It was reported that the acid content of mature yellow carambola fruit averaged 13 mg g^{-1} fresh weight, and the major acid found was oxalic acid followed by tartaric, α-ketoglutaric and fumaric acids (Cambpell & Koch 1987, 1989). For example, the major oxalic acid found, was quantified in 15 carambola cultivars and levels varied almost 10-fold between the different cultivars assessed (Wilson et al. 1982).

The carotenoids content of carambola was also investigated and their levels averaged 22 μg g^{-1} fresh weight. The main pigments identified, were phytofluene (17%), ζ-carotene (25%), β-cryptoflavin (34%) and mutatoxanthin (14%), while-carotene, β-apo-8-carotenal, cryptoxanthin, cryptochrome and lutein were found in small amounts (Gross et al. 1983). On the other hand, Yap et al. (2009) investigated the total phenolic compounds (TPC) of carambola, and they found that TPC ranged from 23.66 to 25.10 mg g^{-1} dry weight. Further studies on phenolic compounds of carambola yielded anthraquinones, flavone, C-glycosides, anthocyanins and dehydroabscisic alcohols. The isolation of other alkyl phenols, 2,5-dimethoxy-3-undecylphenol and 5-methoxy-3-undecylphenol, and two benzoquinones, 5-*O*-methylembelin and 2-dehydroxy-5-*O*-methylembelin was also reported by Araho et al. (2005) and Chakthong et al. (2010).

7.1. Medicinal Uses

Because of its composition and food value, carambola fruit, as well as other parts of the tree, have been used in traditional medicine. The fruit and leaves were used for eczema, digestive aid, and to keep the body cool (Rahmatullah et al. 2010; Jahan et al. 2011). In India, the ripe fruit is administered to halt

Table 5.1. Chemical composition of carambola fruit

	Food value per 100 g of edible portion			
Component	Green*	Half-Ripe*	Full ripe*	Full Ripe**
Calories	35.5	-	36	37
Moisture (g)	95.60	95.90	95.71	90.23
	Food value per 100 g of edible portion			
Component	Green*	Half-Ripe*	Full ripe*	Full Ripe**
Soluble Protein (g)	0.65	0.83	0.83	0.87
Amino acids (g)	0.12	0.16	0.17	0.90
Reducing sugars (g)	0.33	1.15	1.32	-
Total sugars (g)	1.12	1.50	2.25	7.25
Pectins (g)	1.74	1.95	5.11	-
Oxalic acid (g)	0.63	0.85	1.04	-
Total chlorophylls (mg)	4.05	13.60	2.10	-
Ash (g)	-	-	-	0.5
Calcium (mg)	-	-	-	0.9
Phosphorus (mg)	-	-	-	11.1
Iron (mg)	-	-	-	0.06
Carotene (µg)	-	-	-	21
Thiamine (µg)	-	-	-	0.05
Riboflavin (µg)	-	-	-	0.044
Niacin (µg)	-	-	-	0.71
Ascorbic Acid (mg)	-	-	-	35

Source: Patil et al. 2010**; Wenkam & Miller 1965*.

haemorrhages and to relieve bleeding haemorrhoids, and the dried fruit or the juice may be taken to counteract fevers. A conserve (mixed fruit jams) of the fruit is said to allay biliousness and diarrhea and to relieve alcohol "hangover" (Manda et al. 2012). A salve made of the fruit is employed to relieve eye afflictions. The leaves are also used as antipruritic, antipyretic and as treatment for scabies (Chanda et al. 2011).

8. RIPENING AND HARVESTING

During maturation and ripening, the weight of the green mature carambola fruit is different from that of half-ripe and fully-ripe fruit. The pH of the fruit increases with maturity, and ripe fruits are found to be less acidic (pH 3.44) than green mature (pH 2.40) and half-ripe (pH 2.71) fruits. The titrable acidity, reducing sugars and tannin contents are also different in fruit at different stages of maturity (Narain et al. 2001).

Lam and Wan (1983) reported that carambola fruit was determined to be a non-climacteric type fruit (fruit that do not have ethylene crisis to ripen and must harvest ripe), and the characteristic upsurge of carbon dioxide exhibited by climacteric fruit during the ripening process, was not evident in the fruit of different stages of maturity. They also noted that Ethrel (exogenous ethylene) treated fruit, showed neither a sudden sharp increase in carbon dioxide nor an ethylene peak. Later, Mitcham and McDonald (1991) investigated some parameters of carambola fruit during ripening. Harvested fruit at four different stages of ripeness (dark green (DG), light green, colour break (CB) and ripe) showed an increase of the ratio CIE colour a/b ratio during ripening and a decrease in firmness. Respiration and ethylene production of carambola fruit of different ripeness stages, suggested a possible climacteric pattern, although daily monitoring of individual fruit respiration and ethylene production did not provide conclusive evidence of the climacteric/non-climacteric nature of carambola fruit. This evidence is needed to determine the handling and storage conditions of the fruit and determine the optimal storage conditions particularly when using modified atmosphere packaging.

Mitcham and McDonald (1991) also reported that the cell walls of green carambola fruit are comprised mainly of cellulose (60%) and hemicellulose (27%), with pectin polymers accounting for 13%. They noted an increase in the proportion of less tightly bound chelator-soluble pectins and a decrease in covalenty-bound pectins during ripening. It was reported that at early ripening phase, no apparent pectin solubilisation is observed, and the loosely-bound water- and chelator-soluble pectins are the first pectic polysaccharides to be affected. They also noted that a number of cell wall hydrolases are involved in the ripening process of carambola fruit. The predominant hydrolases appear to be β-(1,4)-glucanase (as carboxymethylcellulase), pectinesterase, β-galactosidase, and polygalacturonase (PG). However, the activity of other significant hydrolases increases during ripening such as the glycosidases, α-arabinosidase, α-galactosidase, and α-mannosidase, and also the glycanases, β-(1,4)-galactanase and xylanase. Thus, pectins and hemicelluloses are differentially modified, and the levels of buffered-phenol cell wall materials, total polyuronides as well as arabinose, galactose, xylose, and glucose decreased throughout ripening (Chin et al. 1999; Ali et al. 2004a).

Mohd Zainudin (2014) observed the variations in bioactive compounds and antioxidant activity of carambola (cv. B17) fruit at different ripening stages harvested from week 9 until week 13. Ascorbic acid, total carotenoids content (TCC) and sugar composition were increased while total phenolic content (TPC), total flavonoids content (TFC) and β-carotene showed reversed trends as the ripening process progressed. The tocopherol compounds varied differently with contribution of α- and β-tocopherols highest in week 12 and then decreased in week 13 but γ- and δ-tocopherols progressively decreased during ripening. Meanwhile, antioxidant activities also decreased. Bioactive compounds such as TPC, TFC, β-carotene, γ- and δ-tocopherol were found to be dominant in unripe fruit while those of sugar (sucrose, glucose and fructose), TCC, α- and β-tocopherol were prominent in the ripe fruit.

The fruit of carambola ripens well on the trees, while those harvested before full ripening stage do not ripen well. At full maturity and when fully ripe, fruit naturally fall to the ground, and for adequate handling and marketing, fruit should be harvested manually when they start having a touch of yellow. To evaluate potential maturity indices of carambola fruit, Campbell and Koch (1989) evaluated and examined the changes in sugars and acids. They noted that commercial maturity (colour break) occurred 60 to 65 days after fruit, and was highly variable (51 to 103 mm long) and not a reliable indicator of maturity. However, total soluble sugar concentration, mainly glucose and fructose, was higher, and at harvest sucrose constituted 15 to 20% of the total soluble sugars. Sugar accumulation, acid reduction, and colour development continued for at least 7 days after colour break when fruit remained on trees, but such fruit were not firm enough for typical commercial handling.

Commercially, carambola fruit are harvested by hand when a yellow colour begins to develop called "colour break" in the furrow between the ribs while the tips of the ribs (fins) remain green (Brown et al. 1985), and Brown and Wong (1984) reported that most cultivars must be harvested at this close to full colour-break stage to have maximum sweetness. However, Sargent and Brecht (1990a) reported that ethylene pre-treatment allows early harvesting of the fruit. At this stage, fruit are easily stored or shipped well to distant markets. During harvesting in the field, fruit are generally placed into five to ten kg boxes of different materials (mainly PVC) and should be placed in the shade until transported to the packinghouse for washing, grading, packing, and storage. However, fruit for the fresh market or home use can be left on the tree until fully ripe i.e., yellow to golden yellow in colour (Crane 1994).

Carambola trees have a large range in terms of productivity. Trees that receive adequate horticultural attention can yield between 45 to 113 kg of fruit per tree on average. Yields of up to 136 kg per tree have been reported in some cases and for some cultivars (Campbell & Koch 1987).

9. Storage

In storage, carambola fruit harvested at the "colour-break" stage, have good shelf-life for four weeks at 10°C, three weeks at 15°C, and two weeks at 21°C. However, green and ripe carambola fruit are easily damaged and must be handled with great care, because they bruise easily. Non-ripe fruit should be turned often, until they are yellow in colour and ripe with light brown ribs.

Some studies reported different storage conditions to extend the shelf-life of carambola fruit (see O'Hare 1993). Pale green carambola fruit stored at temperatures ranging between 0 and 21°C showed different behaviour. At 16 and 21°C, necrotic lesions, browning, and shrivelling of the ribs were observed after two to three weeks, respectively. Whilst fruit stored at 10°C remained in a good condition, those stored at 0 and 4°C retained freshness and original appearance for five weeks (Grierson & Vines 1965). Campbell and Koch (1987) and Campbell et al. (1987, 1989) reported similar results when carambola fruit were stored at 5°C for six weeks storage of mature green fruits at 5°C maintained fruit firmness, inhibited the solubilization and depolymerization of chelator-soluble wall polyuronides and reduced the increase in pectinesterase and β-galacto-sidase activities, while at 10°C, the fruit exhibited reduced rates of firmness loss, and associated polyuronide solubilization and depolymerization, as well as the increase in pectinesterase, β-galactosidase, polygalacturonase and β-(1,4)-glucanase activities being markedly retarded compared to fruit stored at 28°C. Storage at both 5 and 10°C also delayed colour change and reduced water loss (Ali et al. 2004a). Pérez-Tello et al. (2001) stored carambola fruit at 2 and 10°C, and 85–90% relative humidity, and noted significant sucrose content increase at 10°C. Kenney and Hull (1986) recommended that optimal shelf-life of carambola fruit is of one and up to six weeks when fruits are stored at 10°C and 7°C, respectively.

As with many other fruit, many changes occur in carambola during storage. After harvesting and during storage, studies reported that sugar concentrations do not vary significantly when fruits are stored at low storage temperatures, but slightly decrease when fruits are stored at 10°C. On the other hand, acidity, mainly oxalic and malic acids, slightly decreases in fruit stored at 10°C, but does not change when fruit are stored at 5°C (Campbell et al. 1987). It is also noted that weight loss and colour development are slowed when fruits are stored at 5°C compared to storage at higher temperatures (Campbell et al. 1987: Siller-Cepeda et al. 2004).

10. POSTHARVEST TECHNOLOGIES

There are several postharvest techniques that can be used to extend the shelf-life of fresh crops, especially sub-tropical and tropical fruits that are more susceptible to deterioration and are more perishable. These techniques help maintain, freshness of fruit harvested at later maturity stages, and postharvest quality attributes of the fruit. Besides refrigeration, modified atmosphere (MA) and edible wax coatings are the most commonly used techniques on carambola fruit to retain water and limit respiration, although other techniques have been tested with encouraging results. During storage of carambola, water loss is the main problem because it leads to a loss in weight and causes shrivelling of the fruit. However, storing carambola fruit in high relative humidity conditions, helps to slow water loss and maintain firmness (Ali et al. 2004b). On the other hand, high humidity storage is impractical in most commercial settings, because it also favours fungal diseases development. Therefore, creating a modified atmosphere by using specific films or using edible coating such as waxing the fruit might be advantageous (Ali et al. 2004b).

Chitosan 0.3%' and gum arabic 1%' coatings maintained the carambola fruit quality during postharvest (Gol et al. 2015). Carambola fruit coated with edible coatings (chitosan, gum arabic, and alginate) resulted in delay in weight loss, decay percentage, titrable acidity, pH, TSS, sugar accumulation, pigment degradation, and ascorbic acid content. The edible coatings showed a positive effect on maintaining higher concentration of total phenolics, which decreased in control (untreated) fruit due to their over-ripening and senescence processes (Gol et al. 2015). In another study, Baraiya et al. (2014) tested 2% sodium alginate + 0.1% olive oil, 2% sodium alginate + 0.2% olive oil, and 2% sodium alginate + 0.1% olive oil + 0.25% GTE coatings and fruit was stored at 25 ± 5°C and 65 ± 5% RH. Weight loss and decay occurrence were least in the fruit treated with 2% sodium alginate + 0.1% olive oil. The composite edible coatings tested in this study enhanced the shelf life of coated carambola fruit.

Ali et al. (2004a) investigated the storage of carambola fruit at 10°C using low density polyethylene MAP. They noted that MAP markedly retarded the decline in tissue firmness and the development of fruit colour, restricted water loss, and suppressed the incidence of chilling injury. The delay of fruit softening by MAP correlated closely with delayed solubilization and depolymerization of the chelator-soluble polyuronides. This positive effect of MAP might partly be attributed to suppression of the increase in activity of the major wall hydrolases. However, they also found that colour development of carambola was more closely linked to the storage temperature than to a modified atmosphere packaging (MAP) treatment (Ali et al. 2004b).

Minimal processing of carambola fruit was also investigated. Results show that sliced fresh fruit have limited marketability due to cut-surface browning. Nevertheless, browning and changes in the composition of sliced and whole carambola fruit during storage at low temperature varied considerably between cultivars but, no difference was observed between fruit harvested at mature green or breaker stages of maturity. Treating sliced fruit with 1.0 or 2.5% citric acid + 0.25% ascorbic acid (in water) prior to packaging was very effective in limiting browning (Weller 1997).

Matthews et al. (1989) investigated the treatment of carambola slices by hot water dip, antimicrobial dip and citric acid dip. Slices were stored in three packaging films with oxygen permeability ranging from 15 to 8000 cc m^{-2} 24 hr^{-1}. They observed excessive browning of slices packed in film with very high (8000 cc m^{-2} 24 hr^{-1}) oxygen permeability, but slices dipped in citric acid and vacuum packaged retained satisfactory colour, texture and flavour for six weeks. They also noted that hot water dipping resulted in darkening of the slices, while Sodium benzoate-potassium sorbate (antimicrobial) dipped slices had reduced yeast and mold counts but lower flavour acceptance.

Results reported by Ding et al. (2007) who studied minimally processed carambola slices from different ripening stages and then dipped in 0, 15 and 30 mg L^{-1} ascorbic acid (AA) showed that AA treatment decreased cut surface browning degree, and flesh colour, sweetness, flavour and overall taste were affected by AA treatment. The authors also noted that skin colour, flesh firmness and vitamin C content, cut surface browning of processed carambola decreased during storage.

Teixeira et al. (2007) evaluated postharvest changes in fresh-cut carambola slices in three different packages [polyethylene terephthalate [PET, Neoform® N94], polystyrene trays covered with PVC 0.017 mm (Vitafilm®, Goodyear) and vacuum sealed polyolefin bags (PLO, Cryovac® PD900)] stored at 6.8°C and 90% RH for 12 days. Results showed that PET trays and PVC film did not modify the internal atmosphere and the high, water permeability of PVC led to more rapid slice desiccation, while PLO vacuum sealed bags, which reduced de-greening and led to better appearance maintenance for up to 12 days. Teixeira et al. (2008) also investigated the effect of ascorbic acid, citric acid and Ca-EDTA, controlled atmosphere (0.4 - 20.3% O_2) and the association of these processes on the browning of fresh-cut carambola fruit. They noted that post-cutting dip and low-oxygen atmospheres did not prevent discoloration. However, ascorbic acid (0.5% and 1%) dips reduced browning during storage at 4.5°C. Although cut-surface browning of slices was not relevant, carambola slices treated with 1% ascorbic acid in association with 0.4% oxygen did not present significant browning or loss of visual quality for up to 12 days, 3 days longer than low oxygen alone (0.4% O_2). The authors concluded that fresh-cut carambola quality can be improved by combining both treatments.

Controlled atmosphere was also used to store carambola fruit. When the fruit is stored in an atmosphere of 2.2 to 4.2% O_2 with 8 to 8.2% CO_2 and a temperature of 7°C, it retained a good colour and firmness compared to fruit stored under air (control) (Revel & Thompson 1994). Wan and Lam (1984) also reported that sealed polyethylene film bags delay de-greening of fruit and did not affect flavour for up to one week at 20°C of either green or full-coloured fruit when the final CO_2 content in the bag ranged from 2.5 to 4.5% and O_2 content was about 15%. The use of high pressures (Hypobaric treatment, HP) to reduce the browning of carambola fruit slices was also assessed. Fruit was sliced, vacuum-sealed, and processed at 600 MPa and 800 MPa (for 1 to 6 min), and stored at 3°C for two and four weeks. The colour was evaluated after air exposure, and all the 800 MPa treatments reduced browning compared to the control, but the 600 MPa treatment was less effective (Boynton et al. 2002).

Ethylene treatment has been used commercially to trigger the de-greening process, to hasten ripening and promote a uniform appearance of fruit including carambola (Oslund & Devenport 1983; Reid 2002). As postharvest treatment, the effectiveness of ethylene in the de-greening treatment of carambola was investigated and the results have been considered encouraging for further investigation (Sargent & Brecht 1990b; Miller & McDonald 1997). The antagonist of ethylene biosynthesis 1-methylcyclopropene (1-MCP) was also tested commercially to delay ripening and senescence in many fruits and vegetables. 1-MCP works by out-competing ethylene to prevent the increase in respiration that is caused by ethylene exposure (Sisler et al. 1996a, 1996b). 1-MCP can maintain the firmness of climacteric and non-climacteric fruit when applied postharvest (Martinez-Romero 2003; Jiang et al. 2004; Bregoli et al. 2005). Few studies have been carried out on the effects of 1-MCP on carambola fruit. Teixeira and Durigan (2006) reported that carambola fruit treated with gaseous 1-MCP for 24 hours maintained good colour, and the treatment decreased fruit respiration, but did not extend the ripening period. Warren (2009) investigated the application of 1-MCP to carambola fruit and showed that treatments were beneficial in maintaining fruit firmness, but did not affect fruit nutritional composition, and concluded that 1-MCP helped to maintain firmness of treated fruit and extended shelf-life, therefore, it could be a valuable treatment for carambola that are harvested at the mid-yellow ripening stage.

Another postharvest technology used to extend the shelf-life of carambola fruit is waxing. The first application of this technology was reported by Vines and Grierson (1966), who reported that waxing of carambola fruit tended to reduce weight loss and to retard the green to yellow colour change. Therefore, if carefully handled to avoid mechanical damage, waxed carambola fruit can be stored and kept in a good condition for 3 to 4 weeks at 10°C. Miller and McDonald (1993) reported similar results and showed that waxing carambola fruit reduced weight loss, and fin browning, but led to surface pitting. Warren (2009) also reported that waxing of carambola reduced weight loss but, affected the colour (chilling injury) even when stored at 5°C and this treatment did not show efficacy in maintaining the postharvest quality attributes of the fruit.

11. POSTHARVEST DISEASES AND DISORDERS

11.1. Disorders

To date, few postharvest physiological disorders of carambola fruit have been reported, except chilling injury (CI). Still, compared to other subtropical and tropical fruits, carambola is not especially sensitive to chilling. However, during storage at low temperature ranging from 0°C to 5°C for two to six weeks, some small surface pitting and rib edge browning can occur, and the severity of injury increases with storage time (Wan & Lam 1984; Warren 2009) (see Figure 5.2).

Ali et al. (2004a) noticed CI symptoms in fruit kept for 10 and 20 days at 5 and 10°C, and the incidence and severity of CI symptoms increased with time of exposure to low temperatures. They also noticed that CI was enhanced by the rate of water loss and colour development, however, MAP storage seemed to contribute to increased tolerance of carambola fruit to CI incidence. According to Pérez-Tello et al. (2001), CI symptoms are associated with the activities of peroxidase (POD) and phenylalanine ammonia-lyase (PAL) enzymes, and storage of carambola fruit at low temperatures (2 and 10°C) induced CI including darkened ribs and skin desiccation.

On the other hand, some authors reported that fruit with sharp rib edges are more susceptible to bruising and discoloration than fruit with rounded rib edges (Watson et al. 1988; Knight 1989). This discoloration is likely related to water loss, and appears to be accelerated by rapid water loss (Brown & Wong, 1985). Although the mechanism is unknown, another disorder of darkening and breakdown of the area between the fruit ribs was reported by Campbell et al. (1985).

Figure 5.2. Internal and external browning of carambola harvested ripe and stored at 5°C for 10 days. Left: cross section of a fruit with arrow indicating areas of chilling injury. Right: fruit with external chilling injury observed.

11.2. Diseases

Diseases are often the most important constraint to the production and storage of tropical fruits. They reduce quality attributes of fruits after harvesting, and cause aesthetic problems that lower the marketability of the harvested produce. Among the different known diseases of tropical fruits, anthracnose (*Colletotrichum gloeosporioides*) is the most commonly observed on carambola fruit, with symptoms being thin, light brown patches on fruit edges (Watson et al. 1988). The loss caused by this disease results from the rotting of the ripe fruit after harvest, and even before. In the early stages, lesions appear as small, circular, slightly sunken spots in the skin, and as these spots increase in size the central portion becomes dark because of the presence of the mycelium just beneath the skin (McMillan 1986).

Brown and Wong (1985) identified different diseases on carambola fruit after harvest, and the major ones isolated from the body lesions were *Colletotrichum acutatum* and *Botrytis* sp., while *Phomopsis* sp. was isolated from the necrotic ridges of the ribs. Rots caused by *Dothiorella* sp. and *Ceratocystis* sp. (Watson et al. 1988) and *Cercospora* sp. (Wan & Lam 1984) were also found on carambola fruit, and the latter was most severe when the calyx was not removed prior to storage.

Although minor compared to those commonly observed in other fruit, many other diseases have been identified on carambola fruit throughout the world, e.g., *Phomopsis* sp., *Penicillium* sp., and *Collelotrichum* sp. (Campbell et al. 1989), *Collelotrichum gloeosporoides* (McMillan, 1986), *Cladosporium cladosporoides*, *Alternaria alernata* and *Khuskia oryzae* (Jain & Saksena 1984). Diseases due to *Alternaria alternata*, *Cladosporium cladosporioides* and *Botryodiplodia theobroma* have also been reported, and these diseases mainly occur at physical injury sites with prolonged storage.

CONCLUSION

Carambola fruit is well known in many countries in particular in the tropics and sub-tropics but simultaneously, much more needs to be known about this fruit and its nutritional and food security potential. Indeed, carambola tree is excellent for home landscaping and in some areas, is mainly used as an ornamental, because of its dark green foliage, attractive lilac flowers and star shaped fruit. The fruit is more often eaten fresh, cut up in fruit salads, or used as a garnish for drinks and cocktails. The juice makes a delicious iced drink alone or in combination with other beverages. Therefore, the last decades have seen the popularity of carambola fruit increasing gradually as more and more people become aware of its exotic taste and nutritional value. The fruit is nowadays produced in large orchards in USA, Israel, India, Malaysia, Australia and many other countries. In the Caribbean, carambola fruit is often eaten fresh because people find it thirst-quenching. However, the nutritional and commercial potential of fruit has been barely tapped and much more work is required on various aspects of its physiology and biochemistry. Furthermore, more research should be carried out on how to develop its agro-processing value-added potential in terms of products such juices, jams, jellies, candied, canned, and many other agro-processed potential carambola products.

REFERENCES

Ali, AM, Chin, LH, & Lazan, H 2004a, 'A comparative study on wall degrading enzymes, pectin modifications and softening during ripening of selected tropical fruits', *Plant Science*, vol. 167, pp. 317–327.

Ali, ZM, Chin, LH, Marimuthu, M, & Lazan, H 2004b, 'Low temperature storage and modified atmosphere packaging of carambola fruit and their effects on ripening related texture changes, wall modification and chilling injury symptoms', *Postharvest Biology and Technology*, vol. 33, pp. 181–192.

Ali, SH and Jaafar, MY 1992, 'Effect of harvest maturity on physical and chemical characteristics of carambola (*Averrhoa carambola* L.)', *New Zealand Journal of Crop and Horticultural* Science, vol. 20, pp. 133–136.

Araho, D, Miyakoshi, M, Chou, WH, Kambara, T, Mizutani, K, & Ikeda, T 2005, 'A new flavone C-glycoside from the leaves of *Averrhoa carambola*. *Natural Medicines*, vol. 59, pp. 113–116.

Baraiya, NS, Ramana, Rao TV, & Thakkar, VR 2014, 'Enhancement of storability and quality maintenance of carambola (*Averrhoa carambola* L.) fruit by using composite edible coating', *Fruits*, vol. 69, pp. 195-205.

Berry, RE, Wagner, CJ, Shaw, PE, & Knight, RJ 1977, 'Promising products from tropical fruits', *Food Products Development*, vol. 7, pp. 109–112.

Boynton, BB., Sims, CA, Sargent, S, Balaban, MO, & Marshall, MR 2002, 'Quality and stability of precut mangos and carambolas subjected to high-pressure processing', *Journal of Food Science*, vol. 67, pp. 409–415.

Bregoli, AM, Ziosi, V, Biondi, S, Rasori, A, Ciccioni, M, Costa, G, & Torrigiani, P 2005, 'Postharvest 1-methylcyclopropene application in ripening control of 'Stark Red Gold' nectarines: temperature-dependant effects on ethylene production and biosynthetic gene expression, fruit quality, and polyamine levels', *Postharvest Biology and Technology*, vol. 37, pp. 111–121.

Brown, BI, and Wong, LS (1985) *Report on postharvest physiological studies of rambutan, carambola and sapodilla.* Queensland Primary Industry Department Note, HPG-A325.

Brown, BI & Wong, LS 1984, '*Report on postharvest physiological studies of rambutan, carambola and sapodilla.* Queensland Primary Industry Department Note, HPG-A324.

Brown, BL, Wong, LS, & Watson, BJ 1985, 'Use of plastic film packaging and low temperature storage for postharvest handling of rambutan, carambola and sapodilla. *Proceedings of the Postharvest Horticultural Workshop*, Melbourne, Australia, pp. 272–286.

Burkill, IH 1966, *A dictionary of the economic products of the Malay peninsula*, Ministry of Agriculture, Kuala Lumpur.

Campbell, CA & Koch, KE 1989, 'Sugar/acid composition and development of sweet and tart carambola fruit', *Journal of the American Society for Horticultural Science*, vol. 114, pp. 455–457.

Campbell, CA & Koch, KE 1987, 'Weight, color and composition of developing carambola fruit', *HortScience*, vol. 22, pp. 18.

Campbell, CA, Knight, RJ, & Olszack, R 1985, 'Carambola production in Florida', *Proceedings of the Florida State Horticultural Society*, vol. 98, pp. 145–149.

Campbell, CA, Huber, DJ, & Koch, KE 1989, 'Postharvest changes in sugars, acids, and color of carambola fruit at various temperatures', *HortScience*, vol. 24, pp. 472–475.

Campbell, CA, Koch, KE, & Huber, DH 1987, 'Postharvest response of carambolas to storage at low temperatures', *Proceedings of the Florida State Horticultural Society*, vol. 100, pp. 272–275.

Campbell CW 1965, 'The Golden Star carambola', *Florida Agricultural Experiment Station Journal Series* S-173.

Chakthong, S, Chiraphan, C, Jundee, C, Chaowalit, P, & Voravu-thikunchai, S 2010, 'Alkyl phenols from the wood of *Averrhoa carambola*', *Chinese Chemical Letters*, vol. 21, pp. 1094–1096.

Chanda, S, Dave, R, & Kaneria, M 2011, '*In vitro* antioxidant property of some Indian medicinal plants', *Research Journal of Medicinal Plants*, vol. 5, pp. 169–179.

Chandler, WH 1958, *Evergreen orchards*, Lea and Febiger Publisher, Philadelphia, PA.

Chin, LH, Ali ZM, & Lazan, H 1999, 'Cell wall modifications, degrading enzymes and softening of carambola fruit during ripening', *Journal of Experimental Botany*, vol. 50, pp. 767–775.

Crane, JH 1994, 'The carambola (starfruit)', *Institute of Food and Agricultural Science University of Florida*, Fact Sheet HS–12.

Ding, P, Ahmad, SH, & Ghazali, HM 2007, 'Changes in selected quality characteristics of minimally processed carambola (*Averrhoa carambola* L.) when treated with ascorbic acid', *Journal of the Science of Food and Agriculture*, vol. 87, pp. 702–709.

Gol, NB, Chaudhari, ML, & Ramana-Rao, TV 2015, 'Effect of edible coatings on quality and shelf life of carambola (*Averrhoa carambola* L.) fruit during storage', *Journal of Food Science and Technology*, vol. 52, pp. 78-91.

Grierson, W & Vines, HM 1965, 'Carambola for potential use in gift fruit shipments', *Proceedings of the Florida State Horticultural Society*, vol. 78, pp. 349–353.

Gross, J, Ikana, R, & Eckhardtb, G 1983, 'Carotenoids of the fruit of *Averrhoa carambola*. *Phytochemistry,* vol. 22, pp. 1479–1481.

Ismail, MR & Noor, KM 1996a, 'Growth and physiological processes of young starfruit (*Averrhoa carambola* L.) plants under soil flooding,' *Scientia Horticulturae*, vol. 65, pp. 229–238.

Ismail, MR & Noor, KM 1996b, 'Growth, water relations and physio-logical processes on starfruit (*Averrhoa carambola* L.) plants under root growth restriction', *Scientia Horticulturae*, vol. 66, pp. 51–58.

Jahan, FI, Hasan, MR, Jahan, R. Seraj, S, Chowdhury, AR, Islam, MT, Khatun, Z, & Rahmatullah, M 2011, 'A comparison of medicinal plant usage by folk medicinal practitioners of two adjoining villages in Lalmonirhat district, Bangladesh', *American-Eurasian Journal of Sustainable Agriculture*, vol. 5, pp. 46–66.

Jain, S and Saksena, SB 1984, 'Three new soft diseases of *Averrhoa carambola* from India', *National Academy of Science Letters of India*, vol. 7, pp. 327–328.

Jiang, W. Zhang, M, He, J, & Zhou, J 2004, 'Regulation of 1-MCP-treaded banana fruit quality by exogenous ethylene and temperature', *Food Science and Technology International,* vol. 10, pp. 15–16.

Kenney, P and Hull, L 1986, 'Effects of storage conditions on carambola quality', *Proceedings of the Florida State Horticultural Society*, vol. 99, pp. 222–224.

Knight, RJ 1965, 'Heterostyly and pollination in carambola', *Proceedings of the Florida State Horticultural Society*, vol. 78, pp. 375–378.

Knight, RJ 1982, 'Partial loss of self-incompatibility in 'Golden Star' carambola', *HortScience*, vol. 17, pp. 72.

Knight, RJ 1989, 'Carambola cultivars and improvement programs', *Proceedings of the Inter-American Society of Tropical Horticulture*, vol. 33, pp. 72–78.

Lam, PF & Wan, CK 1983, 'Climacteric nature of the carambola (*Averrhoa carambola* L.) fruit', *Pertanika Journal of Tropical Agricultural Science*, vol. 6, pp. 44–47.

Manda, H, Vyas, K, Pandya, A, & Singhal, G 2012, 'A complete review on Averrhoa carambola', *World Journal of Pharmacy and Pharmaceutical Sciences*, vol. 1, pp. 17-33.

McMillan, RT 1986, 'Serious diseases of tropical fruits in Florida. *Proceedings of the Florida State Horticultural Society*', vol. 99, pp. 224–227.

Martinez-Romero, D, Dupille, E, Guillen, F, Valverde, JM, Serrano, M, & Valero, D 2003, '1-methylcyclopropene increases storability and shelf life in climacteric and non-climacteric plums', *Journal of Agricultural and Food Chemistry*, vol. 51, pp. 4680–4686.

Matthews, RF, Lindsay, JA, West, PF, & Leinart, A 1989, 'Refrigerated vacuum packaging of carambola slices', *Proceedings of the Florida State Horticultural Society,* vol. 102, pp. 166–169.

Miller, WR & McDonald, RE 1997, 'Carambola quality after ethylene and cold treatments and storage', *HortScience*, vol. 32, pp. 987–989.

Miller, WR & McDonald, RE 1993, 'Quality of cold-treated 'Arkin' carambola coated with wax or plastic film', *Proceedings of the Florida State Horticultural Society*, vol. 106, pp. 234–238.

Mitcham, EJ & McDonald, RE 1991, 'Characterization of the ripening of carambola (*Averrhoa carambola* L.) fruit', *Proceedings of the Florida State Horticultural Society*, vol. 104, pp. 104–108.

Mohd Zainudin, MA, Hamid, AA, Anwar, F, Osman, A, & Saari, N 2014, 'Variation of bioactive compounds and antioxidant activity of carambola (*Averrhoa carambola* L.) fruit at different ripening stages', *Scientia Horticulturae*, vol. 172, pp. 325-331.

Morton, J 1987, 'Carambola', in JF Morton (ed.), *Fruits of warm climates*, Morton Publisher, Miami, FL

Narain, N, Bora, PS, Holschuha, HJ, & Vasconcelos, MADS 2001, Physical and chemical composition of carambola fruit (*Averrhoa carambola* L.) at three stages of maturity', *Ciencia y Tecnologia Alimentaria,* vol. 3, pp. 144–148.

Núñez-Elisa, R & Crane, JH 2000, 'Selective pruning and crop removal increase early-season fruit production of carambola', *Scientia Horticulturae*, vol. 86, pp. 115–126.

O'Hare, TJ 1993, 'Postharvest physiology and storage of carambola (starfruit): a review', *Postharvest Biology and Technology*, vol. 2, pp. 257–267.

Oslund, CR & Davenport, TL 1983, 'Ethylene and carbon dioxide in ripening fruit of *Averrhoa carambola*', *HortScience*, vol. 18, pp. 229–230.

Patil, AG, Patil, DA, Phatak, AV, & Chandra, N 2010, 'Physical and chemical characteristics of carambola (*Averrhoa carambola* L.) fruit at three stages of maturity', *International Journal of Applied Biology and Pharmaceutical Technology*, vol. 1, pp. 624–629.

Pérez-Tello, GO, Silva-Espinoza, BA, Vargas-Arispuro, I, Briceño-Torres, BO, & Martinez-Tellez, MA 2001. Effect of temperature on enzymatic and physiological factors related to chilling injury in carambola fruit (*Averrhoa carambola* L.)', *Biochemical and Biophysical Research Communications*, vol. 287, pp. 846–851.

Popenoe, W 1920, *Manual of tropical and subtropical fruits*, MacMillan, New York, NY.

Purseglove JW 1968, *Tropical crops: Dicotyledons*, Wiley & Sons Publisher, London.

Rahmatullah, M, Khatun, A, Morshed, N, Neogi, PK, Khan, SUA, Hossan, MS, Mahal, MJ, & Jahan, R 2010, 'A randomized survey of medicinal plants used by folk medicinal healers of Sylhet division, Bangladesh', *Advances in Natural and Applied Sciences*, vol. 4, pp. 52–62.

Reid, MS 2002, 'Ethylene in postharvest technology In: Postharvest technology of horticultural crops. 3rd ed. *University of California Agricultural and Natural Resources Communication Services*, CA, 149–162.

Revel, L & Thompson, AK 1994, 'Carambola in controlled atmospheres. *Tropical Fruits Newsletter, Inter-American Institute for Cooperation on Agriculture*, vol. 11, pp. 7.

Ruehle, GD 1958, 'Miscellaneous tropical and subtropical Florida fruits', *Florida Extension Agriculture Service Bulletin*, vol. 156, pp. 30–33.

Salakpetch, S, Turner, DW, & Bell, B 1990, 'The flowering of carambola (*Averrhoa carambola* L.) is more strongly influenced by cultivar and water strees than by diurnal temperature variation and photoperiod,' *Scientia Horticultura*, vol. 43, pp. 83–94.

Sargent, SA & Brecht, JK 1990a, 'Ethylene pretreatment allows early harvest of the fruits', *HortScience*, vol. 25, pp. 1174.

Sargent, SA & Brecht, JK 1990b, '*Carambola degreening study*. Report. Brooks and Son, Inc., Brooks Tropicals, viewed 23 September 2017, http://www.brookstropicals.com.

Shui, G & Leong, LP 2004a, 'Resides from star fruit as valuable source for functional food ingredients and antioxidant nutraceuticals', *Free Radical Biology and Medicine*, vol. 36, pp. S132–S132.

Shui, G & Leong, LP 2004b, 'Analysis of polyphenolic antioxidants in start fruit using liquid chromatography and mass spectrometry', *Journal of Chromatography A*, vol. 1022, pp. 67–75.

Siller-Cepeda, J, Muy-Rangel, D, Baez-Sanudo, M, Garcia-Estrada, R, & Araiza-Lizarde, E. (2004) Quality of carambola (*Averrhoa carambola* L.) fruits harvested at four stages of maturity', *Serie Horticultura*, vol. 10, pp. 23–29.

Sisler, EC, Dupille, E, a&nd Serek, M 1996a, 'Effect of 1-methylcyclopropene and methylenecyclopropane on ethylene binding and ethylene action on cut carnations', *Plant Growth Regulators*, vol. 18, pp. 79–86.

Sisler, EC, Serek, M, & Dupille, E 1996b, 'Comparison of cyclopropene, 1-methylcyclopropene and 3,3-dimethylcyclopropene as ethylene antagonists in plants', *Plan Growth Regulation*, vol. 18, pp. 169–174.

Teixeira, GHA, Durigan, JF, Alves, RE, & O'Hare, TS 2008, 'Response of minimally processed carambola to chemical treatments and low-oxygen atmospheres', *Postharvest Biology and Technology*, vol. 48, pp. 415–422.

Teixeira, GHA, Durigan, JF, Alves, RE, & O'Hare, TS 2007, 'Use of modified atmosphere to extend shelf life of fresh-cut carambola

(*Averrhoa carambola* L. cv. Fwang Tung)', *Postharvest Biology and Technology*, vol. 44, pp. 80–85.

Teixeira, GHA. & Durigan, JF 2006, 'Controle do amadurecimento de carambolas com 1-mcp', ['Control of ripening carambolas with 1-mcp'] *Revista Brasileira de Fruticultura Jaboticabal*, vol. 28, pp. 339–342.

Vines, HM & Gierson, W 1966, 'Handling and physiological studies with carambola', *Proceedings of Florida State Horticultural Society*, vol. 79, pp. 350–355.

Wagner, CJ, Bryan, WL, Berry, RE, & Knight, RJ 1975, 'Carambola selection for commercial production', *Proceedings of the Florida State Horticultural Society*, vol. 88, pp. 466–469.

Wan, CK & Lam, PF 1984, 'Biochemical changes, use of polyethylene bags and chilling injury of carambola (*Averrhoa carambola* L.) stored at various temperatures', *Pertanika Journal of Tropical Agricultural Science*, vol. 7, pp. 39–46.

Warren, O 2009, *'Quality of carambola fruit (Averrhoa carambola L.) as affected by harvest maturity, postharvest wax coating, ethylene and 1-methylcyclopropene'*, MSc thesis, University of Florida.

Watson, BJ, George, AP, Nissen, RJ, & Brown, BI 1988, 'Carambola: a star on horizon', *Queensland Agricultural Journal*, vol. 114, pp. 45–51.

Weller, A, Sims, CA, Matthews, RF, Bates, RP, & Brecht, JK 1997), 'Browning susceptibility and changes in composition during storage of carambola slices', *Journal of Food Science*, vol. 62, pp. 256–260.

Wenkam NS & Miller, CD 1965, 'Composition of Hawaii fruits', *Hawaii Agricultural Experiment Station, College of tropical Agriculture, University of Hawaii*, Bulletin No. 135.

Wilson, CW, Shaw, PE, & Knight, RJ 1982, 'Analysis of oxalic acid in carambola (*Averrhoa carambola* L.) and spinach by high-performance liquid chromatography', *Journal of Agricultural and Food Chemistry*, vol. 30, pp. 1106–1108.

Yap, CF, Ho, CW, Aida, WMW, Chan, SW, Lee, CY, & Leong, YS 2009, 'Optimization of extraction conditions of total phenolic compounds from star fruit (*Averrhoa carambola* L.) residues', *Sains Malaysiana*, vol. 38, pp. 511–520.

In: Agriculture, Food, and Food Security
Editor: Clinton Lloyd Beckford
ISBN: 978-1-53613-483-4
© 2018 Nova Science Publishers, Inc.

Chapter 6

FOOD AND NUTRITION SECURITY THROUGH URBAN FOOD GARDENS: THE ROLES AND POTENTIAL OF COMMUNITY GARDENING

Clinton Beckford and Blessing Igbokwe*
Faculty of Education, University of Windsor, Windsor, Ontario, Canada

1. BACKGROUND

According to the Center for Disease Control and Prevention (CDC) (nd) community gardens serve many functions including opportunities to:

- Eat healthy fresh fruits and vegetables.
- Engage in physical activity, skill building, and creating green space.
- Beautify vacant lots.
- Revitalize communities.
- Revive and beautify public parks.
- Create green rooftops.

---

* Corresponding author email: clinton@uwindsor.ca.

- Decrease violence in some neighborhoods, and improve social well-being through strengthening social connections.
- Promote urban renewal and counteract urban blight

This chapter focuses on the potential of community gardens to enhance food and nutrition security and address obesity and other chronic diseases in low-income urban neighbourhoods by improving access to and consumption of healthful and affordable fresh fruits and vegetables. It should be noted that while community gardens are found in urban, suburban and peri-urban and rural areas, this paper focuses on the community gardens that fit the description of urban food gardens, which are designed as spaces for the small-scale production of food to enhance household food and nutrition security often as part of a larger poverty alleviation strategy. There is a growing phenomenon of urban gardens established by private individuals in their backyards and rooftops using a variety of innovative techniques such as air pruning and rain gutter growing system and other hydroponic techniques. These are an interesting component of urban and peri-urban food culture worthy of more research and exposure but, are not the focus of this paper.

We start with a general discussion that provides some background about the nature of community gardens and community gardening. We then provide a brief history of the development of community gardens before turning to a detailed discussion of the nutritional and health potential of community gardens through increased access to quality fresh fruits and vegetables among low-income urban residents. We present a case study of an urban community garden in Windsor, Ontario, Canada before making some concluding remarks.

2. INTRODUCTION

2.1. Brief History of Community Gardening

Community food gardens are found in urban, suburban, peri-urban, and rural areas but, are most prevalent in the urban core which is populated by many low-income residents (Raja et al. 2008). Their goal may be succinctly summarized as the collective farming of small to medium sized urban spaces to produce fresh, healthy, and affordable fruits and vegetables for the urban poor.

Organized community gardens in Europe and the US can be traced back to the early 18th century but, *community gardening* as a principle and praxis goes back much further to customary cultural food production approaches of indigenous populations around the world. The community gardens of today which are established on public lands often with local government support and cooperation, are far more recent dating back to just after World War 2 with the Victory Garden Program in the USA and Europe and expanding rapidly between the 1960s and 1970s. They are in the process of a major grassroots resurgence presently (Pennsylvania Horticultural Society 2009). According to the American Community Gardening Association (2009) there were more than 18, 000 community gardens in the United States and Canada.

The academic literature on community gardens is dominated by the North American experience, specifically the United States and Canada where they have been growing in prominence over the last 30 years in particular. However, community gardens as they are defined here, do exist in other parts of the world. A more common phenomenon in areas like Europe, Asia, and Africa is the broader concept of urban gardening, of which we consider community gardens to be a subset. There are numerous examples and more literature on small-scale gardening by individuals and families in their backyards, front yards and on their rooftops and terraces. There are also examples of larger gardens in peri-urban areas- the transition area between urban and rural spaces.

Groning (1996) discussed the German *allotment gardens* and the difficulties in getting governments to make more public lands available for garden collectives. Groning chronicles the decline of community gardens starting in the 1960s, as competition for public lands increased with pressure to build more schools, more public parks, more residential communities among other pressing land uses.

Jackson (n.d.) in discussing the rapid global expansion of community gardening traces its growth in the UK back to the 1800s when the Government provided lands for food and ornamental purposes. Similar to the US, they expanded with the Victory Gardens after World War 1 and World War 2. Since then they have been instrumental sources of food for many people.

In Australia, community gardens are also extremely popular (Jackson n.d.). This began at the turn of the 20th Century as a response to food shortages. The popularity observed today started in the 1970s with seminal garden projects in Melbourne and spreading to Sidney and other cities subsequently. Community gardening is viewed as an innovative way to grow food and improve health. Gardening is the focal point but community gardens have become community hubs for a plethora of activities that bring people together and build community.

Community gardens are also typical of rural and urban landscape in Asia and have been famously successful in Cuba where they feed millions of people not just in urban areas but in rural areas as well (Beckford & Campbell 2013).

2.2. Urban Community Gardens

2.2.1. Urban Agriculture

Urban and peri-urban agriculture (UPA) has become an important element in food security strategy globally. UPA has been around and been important at the household level for ages but received institutional and policy attention when the Food and Agricultural Organization (FAO) of the United Nations, in 1994 launched its Special Program for Food Security

(SPFS) to help address global food insecurity and poverty (FAO 2001) which was adopted at the World Food Summit in 1996. UPA was an important component of the SPFS with the goal of improving access to quality food for urban and peri-urban residents (Beckford & Campbell 2013).

Urban and peri-urban agriculture is defined as the production marketing and distribution of agricultural products within city limits (urban) and on the outskirts of cities and towns (peri-urban) (Beckford and Campbell 2013). UPA is diverse in terms of scale and location but includes community gardens; home gardens -in backyards, front yards and on rooftops, terraces and balconies; abandoned land; along river banks; and along railway tracks (FAO cited by Beckford and Campbell 2013). The activities are more expansive in peri-urban areas where livestock, poultry and aqua-culture might be integrated.

The FAO sees UPA as food production by the poor for the poor- an adoptive response by the urban poor to inadequate, unreliable, and irregular access to food (FAO, 2001). Research indicates that the benefits of UPA are enormous and include enhanced food and nutritional security for low income urbanites, employment, income generation (Pasquini & Young 2009), environmental benefits, as well as reducing individual and household vulnerability (Ambrosi-Oji, 2009).

2.2.2. Community Gardens

Community gardens may be seen as a component of urban agriculture, community gardens are part of a growing trend as a strategy to increase community-wide fruit and vegetable access, availability and consumption (McCormack et al. 2010). Larson et al. (2009) attributes the growing popularity of community gardens largely to their potential to increase fruit and vegetable availability and consequently consumption among low-income residents who live in food deserts. McCormack et al. (2010) see community gardens as having the potential to improve nutrition outcomes, including nutrition knowledge, attitude, and dietary intake. Jackson (n.d.) asserts that community gardening is an innovative way to grow food and improve health.

The American Community Garden Association (ACGA) describes a community garden as a piece of land that is collectively gardened by a group of people (2007). Jackson (n.d.) describes community gardens as spaces where people come together to inter alia, grow food, foster good health, and green urban environments. Typically, they are established on publicly owned lands (Ferris et al. 2001) or land donated by organizations for the expressed purpose of collective gardening. They can take various forms in terms of their purpose and functions. In developed countries in Europe and North America, they might be discrete areas with a defined boundary even without a fence, divided into small plots where people grow small plots of vegetables, fruits and legumes or they might be much larger areas designed as 'green' spaces. In some cities of the developing world such as Dar Es Salaam and Moshi in Tanzania, they are predominantly food gardens established on small pieces of land on roadsides. The latter, are different from community gardens in let's say the United States and Canada, in that they are typically operated by individuals, rather than collectives.

Some more famous community gardens are more sophisticated, with designs inspired by famous architects such as the Clinton Community Garden in Manhattan, New York City and the Peralta Garden in Berkley, California (ACGA, 2007). These have become popular places where food production is only one aspect of a larger healthy community vision that includes art, social interaction, and ecological awareness. They represent success for a paradigm that advocates for reasserting the integrity of *the commons*. These areas are treated as 'green' sanctuaries in large 'concrete jungles' and contribute greatly to the vibrancy of the immediate and surrounding neighbourhoods and communities. In Australia, community gardens have become community hubs for a variety of activities including learning and education, art and creativity, community events, celebrations and social enterprise (Jackson n.d.).

In some areas, community garden movements, projects or collectives, are deliberately intended to boost food and nutrition security in urban and peri-urban areas. For example, the Gardens Project in Lake County and Mendocino County, California, seeks to relieve hunger and inadequate nutrition in low-income neighborhoods, senior communities, schools and youth enterprise projects by creating access to community-based food production and local, nutritious food (Logsdon & Ryan, 2017). The organization does so through a network of activities including school nutrition programs, school and community gardens, food stamp program and seeking to impact food policy.

The produce from community gardens, is typically for personal household use. In some jurisdictions community gardeners who are part of an organization on public lands may even be prohibited from selling the produce from their garden plots. Some gardens require some produce to be donated to local food banks, soup kitchens, community food pantries, homeless shelters, churches and other community social programs. Many gardeners also voluntarily share and donate fresh food with family and relatives, friends and neighbours and community social programs. Some community gardens are also tied to local famers' markets which have been common in many developed countries for some time now and are growing in prominence in both developed and developing countries.

3. FOOD, AGRICULTURE, AND URBAN COMMUNITY FOOD GARDENS

Harris (2009) discussed the potential of community gardens to help in building resilience to some of the expected effects of climate change on food and nutrition security. As environments change in response to change and variability in climate, it is expected that some areas will experience declining food production and reduced availability of affordable fresh produce. Community gardens play an important role in promoting food sovereignty, an approach to food security which has at its core, a commitment to the right of local people to determine their own food futures through increased local production and reduction of reliance on imported food. Urban community gardens allow residents to grow some of their own food which may then be shared and donated with food insecure urban population through the urban community networks that are facilitated (Harris 2009; Nelson 1996). It has also been suggested that locally produced food reduces the use of fuel used to transport food in from outside therefore, benefitting the environment (Kishler 2010, 2012).

Urban community gardens bring participants closer to the source of their food (Beckford and Campbell 2013) and also increase knowledge about food and agriculture as gardeners share their knowledge and expertise about growing food (Jackson n.d.). In many areas, community gardens are multicultural in terms of the garden membership (Aliamo et al. 2008; Eizenberg 2012) leading to significant cultural exchange of food knowledge. Community gardens can also play a big role in food and nutrition security in many large cities as they donate to food banks and soup kitchens. This is an area that may be underestimated because of lack of reliable data and is therefore worthy of rigorous research.

Urban community gardens may also stimulate interest in local food production in general. Some cities have reported an increase in the public's enthusiasm for local urban-agricultural spaces (Wang, Qju & Swallow 2014). In some areas, community gardens have been used as sites for engaging inner-city youth and children about the value growing their own food while addressing food security problems that these communities and households often experience (Lautenschlager & Smith 2007; Morris & Zidenberg-Cherr 2002). Research shows that youth who were involved in an urban gardening program in the United States, demonstrated higher leadership skills and better dietary behaviours than their peers who did not have this experience (Morris et al. 2002; Morris et al. 2002; Morris & Zidenberg-Cherr 2002).

Community gardens can help to advance the growing food sovereignty and food justice agendas in many countries. Local people have very little control or autonomy over their food- where it comes from and how it is produced. The community garden movement democratizes food systems by relocalizing production to meet nutritional and safety needs in accordance with social and cultural values (Hamm & Bellows 2003).

3.1. Who are Community Gardeners?

Urban gardeners generally speaking are diverse in ethic and socio-economic backgrounds. There are people of all race and class engaged in the various forms and scale of urban food cultivation. However, a different picture emerges when we examine community gardening more specifically. What we see is that in many urban areas, community gardeners are mainly people who are economically disadvantaged and in the United States and Canada, racial minorities and immigrants make up a large proportion of urban community gardeners. In one US study, only 26.4% of the community gardeners were white, 61.5% were African Americans and 12% were other races (Aliamo et al. 2008). Eizenberg, (2012) claims that since the 1970s, community gardens in New York City have been managed by, and served the city's economically disadvantaged, especially African Americans and Hispanic residents. They are predominantly located in areas with an average of over 80% African Americans and Hispanics with a household median income of $25, 000.00 compared to the median of $38, 515 in 2000 (Eizenberg 2012).

In terms of socio-economic demographics, community gardens are mainly found in areas that ascribe to the definition of food deserts and so are cultivated mainly by low-income residents. Both male and female participants are involved and there is no data that establishes dominance of any gender. Often, garden plots are cultivated by families or household members with persons of different ages involved (See the case study in this chapter).

3.2. Increased Intake of Fruits and Vegetables and Dietary Benefits

There is a growing body of research about community gardens. Much of this research has focused on the role of these urban spaces in community development. Their roles and functions in terms of dietary intake had not been researched much until fairly recently (Alaimo et al. 2008).

Access to fresh and better tasting food has been regularly identified as a main reason for participation in community gardening (Anderson 2000). The provision of food for low-income households, is seen as one of the significant benefits of community gardening.

The rise in the prevalence of chronic diet related diseases especially in urban populations, has fueled interest in urban community gardening. There is evidence that higher intakes of fruits and vegetables can reduce the risks of cardio-vascular disease, stroke and cancer and numerous other chronic diseases (Bazzano, He & Ogden 2002; McCormack et al. 2010; Riboli & Norat 2013). Research also shows that the lack of access to fruits and vegetables can adversely affect susceptibility to diseases (Bruner et al. 2008; McPhail, Chapman & Beagan 2013). This has led to a desperate search for solutions of which community gardens are seen as a viable option to increase availability of affordable fresh fruits and vegetables (Aliamo et al. 2008; McCormack et al., 2010; Tong, Ren & Mack 2011). In addition, community gardens can reduce health risks by contributing to a healthier dietary pattern (Alaimo et al. 2008; Guitart, Pickering & Bryne 2013; Jerme & Wakefield 2013).

More importantly, it is felt that the intake of fruits and vegetables among the urban poor can be positively affected through engagement in urban community food gardens. This is particularly important in areas with limited access to regular grocery stores like supermarkets (Zenk et al. 2006; Zenk, Shultz & Hollis-Neely 2005; Morland, Wing & Roux 2002) and which have the characteristics of food deserts. The term food desert is used to describe a situation where residents and consumers have limited access to affordable nutritious food. These are typically low-income neighbourhoods with inadequate supermarket facilities (Wang et al. 2014).

A food desert is defined as a 'low access community' where at least 500 people or at least 33% of the population live at least a mile from a supermarket or large grocery store (USAID 2009) and incidentally, they are found in rural as well as urban areas though urban food deserts have been more researched. Corrigan (2011) notes that food deserts are characterized by an abundance of fast food eateries and a limited number of supermarkets or grocery stores offering high quality food. This situation is exacerbated by the fact that the menus of restaurants in local-income neighbourhoods have fewer healthy options than those in high-income areas (Lewis et al. 2005). This helps to explain the rise in obesity among some food insecure segments of urban populations. Food deserts are also characterized by high unemployment and underemployment, preponderance of low-wage jobs, and mobility issues in relation to public transportation to access reliable food sources. People who live in food deserts might not necessarily be hungry. Some may actually be able to obtain food. However, many of these residents are forced to use arrangements such as food banks, food stamps and other social welfare programs and community resources. We submit therefore, that the ability to access food in dignified and legal ways should be an element in the analysis and identification of food deserts and food security more generally.

Studies show that community gardeners are more likely to eat more fruits and vegetables and consume less processed sugar foods and also less milk products (Blair, Giesecke & Sherman 1991). Other studies, show that on average, where a household member was involved in community gardening, there was consumption of fruits and vegetables 4.4 times were day compared to 3.3 for respondents without a community gardener in the household (Alaimo et al. 2008). They concluded that community gardens can improve dietary fortunes of urban people as a household's participation in a community garden was positively associated with increased consumption of fruit and vegetable.

Advocates of community gardens suggest that these food gardens can help to alleviate one of the most critical barriers to the urban poor having a healthy diet, that is, limited availability of fresh produce (Alaimo et al. 2008). The issue of limited availability of fresh produce is no doubt serious. Particularly in the context of food desert areas where there is a predominance of convenience stores and corner shops and limited numbers of supermarkets. In the United States, studies again show that supermarkets are less common in low income neighbourhoods with predominantly racial minorities and further, that presence of supermarkets is associated with higher consumption of fruit and vegetable among minority residents (Morland et al. 2002; Morland, Wing & Roux 2002). Corrigan (2011) suggests that in an area dominated by corner shops that offer limited fresh foods, a community garden may provide an opportunity increase food knowledge and even improve food security.

Research shows that youth who participated in school gardening programs exhibit greater preference for vegetables than those who did not (Allen et al. 2008; Morris et al. 2002; Morris, Neustader & Zidenberg-Cherr 2001; Morris & Zidenberg-Cherr 2002). There is also evidence that youth engagement in community gardening, strengthened their understanding of food systems, the gardening process, and healthy versus unhealthy foods (Lautenschlager & Smith 2007) which would have an influence on their food choices and healthy lifestyles (Corrigan 2011). Studies also show that urban residents want to eat well but are constrained by structural barriers. For example, economically challenged urban women understood the benefits of healthy eating and the nutritional value of fruits and vegetables, but were unable to access them on a regular basis because financial constraints (Eikenberry& Smith 2004; Dibsdall et al. 2003).

Community gardens have the potential to be quite productive, producing large volume of food with relatively few resources (Alaimo et al. 2008). Studies have shown that more farmers' markets and supermarkets per capita, correlates with lower rates of obesity (Leone et al. 2012; Payne et al. 2013). It is logical then to hypothesize that community gardens with their ability to provide fresh fruits and vegetables with low resource inputs, could be part of the solution to food desert problems (Campbell 2012; Guitart et al. 2013).

Wang et al. (2014) in their Canadian study of Edmonton, Alberta, show that community gardens can alleviate the problems of food deserts in neighbourhoods with low accessibility, low median income, low car access and high population density by being alternative fresh food providers. They are most effective at enhancing food security at the individual and household levels and have very limited ability to alleviate food deficits in whole cities. The role they play in this regard is perhaps understated as their contributions to food banks, soup kitchens, meals-on- wheels programs, food pantries, and other social welfare programs and organization have not been the subject of empirical research. Corrigan (2011) posits that donations of food from community gardens is considerable and impactful to the community.

In the context of a food security framework, community gardens can play a significant role. According to USAID (2009) critical dimensions of food security include the availability of enough *nutritious* and *safe* food, and the *ability to procure* this food in socially acceptable ways- which means without using emergency food programs, scavenging, stealing, etc. Community gardening satisfy these dimensions directly. First, the focus of gardens is usually the cultivation of fruits, vegetables and legumes, which are all nutritious and all associated with greater nutrition and health outcomes. Then there is the issue of 'safe' food, a dimension that is often overlooked when food security is discussed. Food from community gardens is considered to be safe as many gardens prohibit the use of harmful chemical inputs such as pesticides and insecticides. Because of this, food from community gardens are often described as organic. Lyson (2004) explained that community gardens allow gardeners to grow and supply their own food in a sustainable way thus facilitating consumption of healthier food. Community gardeners say that the produce from their gardens taste better than store bought food. Gardeners also have more trust in their garden food because they grew it themselves and they know how it was grown and don't have to worry about the problems that make people skeptical of mass produced food transported in from other regions and countries. Baker (2004) posited that vegetables grown in community gardens were five times the national standard in Canada.

Community gardens also meet the critically important dimension of food security-*access*. Access refers to the ability of people to procure food. This relates to whether or not food is affordable-a function of individual or household income. Well-stocked supermarket shelves mean nothing to low-income individuals if they are not able to buy the foods on those shelves anyway. The shelves may as well be empty to those people. This issue of access/affordability is an important aspect of food security. Nobel Prize laureate Amartya Sen, revolutionized our understanding of food insecurity in the 1980s with his explanation of hunger even during times of plentiful food. Sen (1981) couched the term 'entitlements' arguing that famine was caused not by declining availability of food but, by declining entitlements. This leads to inequitable distribution of food of which food insecurity is a direct consequence (Sen 1981). People who lack financial resources are unable to, on their own, afford good quality food (Corrigan 2011). Community gardens address this issue in that people are able to cultivate some of their own food at low cost and avoid the challenges associated with purchasing expensive healthy food or purchasing unhealthy (not necessarily inexpensive) fast-food. Community gardens are not able to singlehandedly resolve the problems of food insecurity and food deserts, but they can enhance local level access to fresh foods (Corrigan 2011).

Community gardens have become an integral component of local food-based systems which include farmers' markets, and backyard urban agriculture. These have become important resources in the quest for urban community food and nutrition security. They are also part of what some describe as 'civic agriculture' which is described as a locally-based agricultural and food production system associated with community development (Lyson 2004).

3.3. Productivity of Community Gardens

Community gardens vary in size but plots are generally fairly small (see case study later in this chapter). This can create an impression that they are not able to significantly impact a person's or a family's food security. However, as we discuss in the case study, these small plots can produce substantial amounts of food. Patel (1991) found that 405 community gardens in Newark, New Jersey produced some $450, 000.00 worth of food in 1989 and that this led to a significant reduction in the food bill of gardeners. A youth garden in Berkeley California sold over $10, 000.00 of food in one year (Lawson 2005).

Many gardeners participate in community gardens because it significantly augmented their food supply (Armstrong 2000). Some gardeners saw community gardening as a substitute for store bought food that had a substantial impact on their household budget (Wakefield et al. 2007) through reduced food costs. Jackson (n.d.) also showed that community garden in Australia was associated with reduced budgetary spending on food among gardeners. Anderson (2000) cites research about one community garden project which estimated savings of $50-$250 per season in food costs for gardeners. It was estimated also that 37 gardeners grew 5000 lbs. of vegetables (Anderson 2000). Community gardeners described growing their own food as a socio-economic benefit to their community. They help to improve household food security of the gardener but also her/his neighbours with whom they routinely share and exchange food. Anderson (2000) reported on research from the United States where 37 gardeners in one project donated 1000 lbs. of vegetables to friends, neighbours, and local community feeding programs and organizations,

Apart from fresh fruits and vegetables, community gardens enhance food security well beyond the period of harvest. Corrigan (2011) found that many gardeners engaged in some amount of small-scale household level agro-processing, specifically food preservation. Many also reported that they simply freeze some food which can last for months on end. Donating food and giving back to the community was a major theme raised in a study of a community garden in Baltimore, Maryland. The Duncan Street Miracle Garden, produces an 'enormous amount of food' and also donates 'an enormous amount...' (Corrigan 2011 p. 1238).

4. Case Study: The Shirley and Ray Gould's Community Garden

The Shirley and Ray Gould Community Garden started in 2013 on land donated by Ford Motor Company. The garden is situated in Windsor, Ontario in Canada and is located at the Unemployment Help Center compound in the east end of the city. The garden is part of the efforts of the Unemployment Help Center to engage local communities and integrate new comers to Canada. The garden is located in a neighbourhood of a large new comer population and the garden plots are cultivated mainly by new comer families. There is a large element of family/household run plots in the garden. During the data collection process, several generations of family members were observed working in the garden. The demographics displayed diversity in age, sex, and to a lesser extent race.

In 2017, the community garden had 106 families registered. The vast majority are Nepalese immigrants but anyone can sign up to join the garden. Conventional garden plots are small- 10x24 feet strips for a total of 240sq.ft. See Figure 6.1. Tools and planting stock are provided by the Unemployment Help Center based on the requests of gardeners. They also provided water and other inputs such as compost and water for irrigation. The Unemployment Help Center also runs an onsite food bank, and each gardener is required to donate to the food bank.

Garden plots are small but produce a surprisingly large volume of food. Cropping systems are akin to subsistence agricultural techniques in the tropics and sub-tropics and among small-scale food farmers and backyard gardeners across the world. Plots are all intensively cultivated with mixed and multiple cropping ta characteristic of almost all the plots. The cropping systems provide a look into the food and agricultural culture and background of these new Canadians. The crops grown are a good indicator of the origin of the gardeners. For example, among Nepalese, the main crops grown are a diverse range of fruit, vegetables, herbs and spices. Gardeners from Central America planted a lot of peas, beans and corn.

Figure 6.1. Conventional garden plots.

Food and Nutrition Security through Urban Food Gardens

Figure 6.2a. Intricate method of growing corn.

Figure 6.2b. Flower pot gardening in open field.

An interesting array of cultivation techniques was observed, which again reflected the origin of the gardeners and showcased the local agricultural knowledge which they had transferred to Canada and applied to their gardening. See Figures 6.2a and 6.2b.

An astounding variety of crops were grown in this small garden. The crops that were identified and recorded included:

1. Sweet Basil
2. Oregano
3. Spearmint
4. Sweet Peppers
5. Lettuce
6. Swiss Chard
7. Tomato
8. Bitter Melon
9. Green Onions
10. Egg Plant
11. Broad Beans
12. String Beans
13. Kidney Beans
14. Cabbage
15. Cucumber
16. Bottle Gourd
17. Squash
18. Okra
19. Beet
20. Onion
21. Sage
22. Tomatillo
23. Leek
24. Carrots
25. Zucchini
26. Chives
27. Foo Gwa
28. Irish Potato
29. Strawberry
30. Celery
31. Corn
32. Summer Savory
33. Rosemary
34. Garlic

Food and Nutrition Security through Urban Food Gardens

These were just some of the main crops with various sub-species of many found. Examples include several types of squash, many types of pepper, tomatoes, okra, cabbage, and corn among other crops. Also of great importance to gardeners, is the ability to produce plants for nutritional and medicinal purposes. Many plants in the garden, served dual purposes-food and nutrition, and medicinal- and were deliberately selected with this in mind. This was of great value to the immigrant families as it was a way to keep connected with some traditional cultural values but, more importantly it provided access to home remedies and cures not available for purchase.

A number of sustainable agricultural techniques were observed as well. One was the individual watering of plants by hand. Another was the planting of Marigold flowers to mitigate against insect pests. A third was the use of straw in one plot to reduce moisture loss and keep down weeds (See Figure 6.3a and 6.3b). A number of techniques are used to grow crops including raised beds, low beds, pots, and raised containers for people with mobility issues (See Figure 6.2b and Figure 6.4).

Figure 6.3a. Ground cover of straw.

Figure 6.3b. Individual watering of plants.

Figure 6.4. Raised beds in containers.

Gardeners share food with each other. They also share knowledge of foods and farming. Some women gardeners spoke about sharing recipes and food preparation and preservation techniques. In some community gardens people participate to network and socialize, get exercise and engage in environmental enhancement.

In this garden, the people were in it for the food and nutrition function. The following statements from gardeners who were interviewed make this point:

> 'The garden allows me to eat the good foods I used to eat in my country'

Another participant when answering the question 'How do you benefit from your garden?' said

> 'I am able to grow some of the food that I need for my family. If I had to buy these products from the superstore, it would be a lot of money.

Yet another person said

> I get a lot of food from my plot even though it is not a lot of land. This was a good season and my family will eat a lot of fresh food. We will have some preserved foods for later too.
>
> He continued to say that 'The food from this garden or my backyard garden is better than supermarket food because we don't use chemicals like the big farms'
>
> A female gardener said 'We should [a] big garden land to grow much food. Then I would not buy vegetables from the store.'

The small plots in the garden are surprisingly high yielding. A main reason for this is that as was alluded to earlier, they are very intensively cultivated. It was difficult to obtain clear data on volume of food produced or reaped because of communication barriers. However, the following account by a female gardener who we were able to communicate shed some light on this issue. She explained that she planted cabbage along the perimeter of the two rows that comprised her plot. She counted 41 plants that reached maturity. She estimated that she obtained approximately 100lbs of produce. She also reaped several dozens of tomatoes, and a large quantity of beans, egg plants and peppers.

Gardeners also spoke about other benefits of gardening such as exercise, relaxation, and meeting people, however, the food and nutrition goal is the fundamental imperative for these gardeners. The garden supplies fresh vegetables but feed gardeners well beyond the period of harvest. Gardeners reported that they are able to preserve certain vegetables for later use and the home remedies they make from medicinal plants can last for months sometimes years without losing their potency.

In addition to the plots that are cultivated by families and individuals, the garden integrates a fruit orchard with an assortment of fruit trees planted along the perimeter of the garden. This include apples, pears, nectarine, peaches and cherries. There are also a few picnic tables strategically located and these will expand when the fruit trees mature and provide more shade.

5. Conclusion

The main problem of food insecurity is one of access; supermarkets believe that suburban shopping centers are safer investments than low-income communities (Flachs 2010) and this is reflected in the location of these facilities and ultimately access to them by low-income urbanites. Financial constraints, or reduction in entitlements (Sen 1981) often lead to reduced food intake and disruption in eating (Office of Nutrition Policy and Promotion-Health Canada cited by Loopstra & Tarasuk 2013). This affects household income and therefore household food security, which is related to heightened nutritional vulnerability and poor health outcomes. Nutritious and fresh food is expensive and difficult to obtain in urban areas, but in areas with a high immigrant or 'ethnic' population, this problem is compounded by people often having limited access to their cultural food traditions.

There is abundant research establishing the association between increased access to fruit and vegetables and increased consumption of fruits and vegetables as well as an association between increased consumption of fruit and vegetables and positive health indices. Evidence from research studies proves conclusively, that community gardens can increase availability and access to fruits and vegetables in low-income areas and food deserts. The location of community gardens in these areas can lead to increased consumption of fruits and vegetables at the individual and household levels, thereby promoting healthy dietary attitudes and behaviour and reducing health risks associated with cardiovascular diseases and other chronic diseases like obesity.

Community gardeners consumed more fruits and vegetables compared with home gardeners and non-gardeners. Their potential make them a unique intervention that can narrow the divide between people and their food and increase local opportunities to eat healthier (Litt et al., 2011).

Community gardens have been successful in mobilizing low-income urban neighborhoods to improve access to fresh produce, engage youth, and improved nutrition by enhancing their access to and consumption of healthy foods (Allen et al., 2008). Flachs (2010) shows how having a hand in producing one's own food creates a tangible connection to produce which make people more invested in their food choices. Studies show that community gardeners are more likely to choose fresh and healthy food over fast and non-nutritious food, all other things being equal. Community gardens create access to affordable, nutritious, and healthy foods particularly vegetables and fruits in food deserts and exert varying levels of influence on the food bill of participants.

Enhancing health and nutritional outcomes for low-income residents through the provision of fresh fruits and vegetables is considered to be one of the major contributions of urban community gardens. These gardens have been associated with increased fruit and vegetable intake among gardeners and gardening families. This food is generally considered to be more nutritious and safe as harmful pesticides and other chemicals are usually prohibited. Many gardens use organic and environmentally sustainable techniques, so that food can be eaten fresh from the soil (Flachs 2010). In a context where people are becoming more conscious of the quality and source of their food, and skepticism about store foods that are labeled 'organic', community gardens are more likely than not to be actual sites of organic farming and other environmentally sustainable gardening and food production practices. Jackson (nd) suggest that a major reason for the interest and surge in community gardening in Australia, is a growing passion for fresh organic produce. This is an important point to make as organic foods in supermarkets are generally very expensive. The People who eat community garden vegetables and fruits are able to eat this healthy food at a minute fraction of the price paid by people who shop in regular grocery stores.

The popularity, demand for, and effects of community gardens, will only continue to increase as solutions are sought for providing more healthy food and an active lifestyle to people and especially those who live in low-income urban areas and food deserts. It can be expected that more and more people including higher-income urban and sub-urban residents, could become involved as consumers become more conscious of, and concerned about their food-where it comes from, how it is produced, and how that impact the environment and sustainability. Research indicates that there is a growing sentiment among consumers globally, that fast food and industrial production options are quite often unhealthy, in addition to being detrimental environmentally and socially. This evolving and expanding process of *food politics* (Flachs 2010), will likely stimulate more interest and participation in local community-based food production and access systems like community gardens, Community Supported Agriculture (CSAs) and farmers' markets. These food systems keep money in the community and engage in socially and environmentally sustainable practices that produce safe, healthy, and affordable foods.

There is a lack of research and scientific literature about community gardening in the developing world and especially in the tropics and sub-tropics. There has been some research about urban agriculture in general but, the concept of community gardens as operationalized here, is very uncommon in these areas many of which are generally food deficit areas or have significant food security challenges. This is perhaps a reflection of cultural and historical norms related to food security specifically and poverty alleviation more generally.

We argue that promoting community gardening as part of a larger urban poverty strategy anchored in a robust food security strategy can bring economic, environmental and health benefits to urban households and communities. This requires rethinking urban land use in some cases, and the use of publicly owned open spaces. Institutionalizing this approach through urban planning policies is an imperative.

ACKNOWLEDGMENT

Research on community gardens for this paper was funded by the University of Windsor, Social Sciences and Humanities Research Grant. We express sincere thanks to the University of Windsor.

REFERENCES

Alaimo, K., Packnett, E., Miles, R. A. & Kruger, D. (2008). 'Fruit and vegetable intake among urban community gardeners' *Journal of Nutrition Education and Behaviour*, vol. *40*, no. 2, pp. 94-101.

Allen, O., Alaimo, J. K., Elam, D. & Perry, E. (2008). 'Growing Vegetables and Values: Benefits of Neighbourhood-based Community Gardens for Youth Development and Nutrition', *Journal of Hunger & Environmental Nutrition*, vol. 3, no. 4, pp. 418-439.

Ambrose-Oji, A. (2009). 'Urban food systems and African indigenous vegetables: defining the spa ces and places for African indigenous vegetables in urban and peri-urban agriculture', in CM. Shackleton, MW Pasquani & AW Drescher, (eds.) *African Indigenous Vegetables in Urban Agriculture*, pp. 1-33, Earthscan, London.

Armstrong, D. (2000). 'A survey of community gardens in upstate New York: Implications for health promotion and community development', *Health and Place*, vol. 6, pp. 319-327.

Baker, L. E. (2004). 'Trending cultural landscapes and food citizenship in Toronto's community gardens', *The Geographical Review*, vol. *94*, no. 3, pp. 305-325.

Bazzano, L. A., He, J., Ogden, L. G., Loria, C. M., Vupputuri, S., Myers, L. & Whelton, P. K. (2002). 'Fruit and vegetable intake and the risk of cardiovascular disease in US adults: The first National Health and Nutrition Examination Survey Epidemiologic Follow-up Study', *American Journal of Clinical Nutrition*, vol. *76*, pp. 93-99.

Beckford, C. L. & Campbell, D. R. (2013). *Domestic Food Production and Food Security in the Caribbean: Building Capacity in Local Food Production Systems*. Palgrave Macmillan, New York, NY.

Bruner, M. W., Lawson, J., Pickett, W., Boyce, W. & Janssen, I. (2008). 'Rural Canadian adolescents are more likely to be obese compared to urban adolescents', *International Journal of Pediatric Obesity*, no. *3*, pp. 205-211.

Campbell, J. N. M. (2012). *'New urbanism and brownfields redevelopment: Complications and public health benefits of brownfield reuse as a community garden'*, Public Health Thesis, Paper 219.

Center for Disease Control and Prevention (CDC) (nd) 'Community gardens'. Available from: https://www.cdc.gov/healthyplaces/healthtopics/healthyfood/community.htm (October 27, 2017).

Corrigan, M. P. (2011). 'Growing what you eat: developing community gardens in Baltimore, Maryland', *Applied Geography*, vol. *31*, pp. 1232-1241.

Dibsdall, L. A., Lambert, N., Bobbin, F. R. & Frewer, J. l. (2003). 'Low-income consumers' attitudes and behaviour towards access, availability and motivation to eat fruit and vegetables', *Public Health Nutrition*, vol. *6*, pp. 159—168.

Eikenberry, N. & Smith, C. (2004). 'Healthful eating: Perceptions, motivations, barriers, and promotors in low-income Minnesota communities', *Journal of the American Dieticians Association*, vol. *104*, pp. 1158-1161.

Eizenberg, E. (2012). 'The changing meaning of community space: Two models of NGO management of community gardens in New York City', *International Journal of Urban and Regional Research*, vol. *36*, no. 1, pp. 106-120.

Flachs, A. (2010). 'Food for Thought: The Social Impact of Community Gardens in the Greater Cleveland Area', *Environmental Green Journal*, vol. *1*, no. 30, Spring 2010, ISSN: 1076-7975. Available from: http://escholarship.org/uc/item/6bh7j4z4 (October 27, 2017).

Food and Agricultural Organization (FAO). (2001). 'The Special Program for Food Security. Urban and Peri-urban Agriculture: A Briefing Guide for the Successful Implementation of Urban and Peri-urban Agriculture in Developing Countries of Transition'. *FAO*, Rome.

Groning, G. (1998). 'Politics of Community Gardens in Germany'. Paper presented at the *1996 Annual Conference of The American Community Gardening Association (ACGA)* "Branching Out: Linking Communities Through Gardening" September 26 - 29, 1996, Montréal, Canada. Published by City Farmer, Canada's Office of Urban Agriculture.

Guitart, D. A., Pickering, C. M. & Bryne, J. A. (2013). 'Color me healthy: Food diversity in school community gardens in two rapidly urbanizing Australian cities', *Health and Place*, vol. *26*, pp. 110-117.

Hamm, M. W. & Bellows, A. C. (2003). 'Community food security and nutrition educators.' *Journal of Nutrition Education and Behaviour*, vol. *35*, no. 1, pp. 37-43.

Harris, E. (2009). 'The role of community gardens in creating healthy communities', *Australian Planner*, vol. *46*, no. 2, pp. 24-27.

Jackson, T. nd, 'The community garden movement', *Foodwise*. Available from: http:// www.foodwise.com.au/ the-community-gardening-movement/ (October 27, 2017).

Kishler, L. (2012). 'Occupy community gardens', *Nation of Change*, February 12, 2012. Available from: http://www.nationofchange.org/occupy-communitygardens-1328107081 (February 2, 2016).

Kishler, L. (2010). 'Community gardens are a serious answer to food supplies, health', *San Jose Mercury News*, March 18, 2010. Available from: http://www.mercurynews.com/opinions/ci_14687546 (February 2, 2016).

Lautenschlager, L. & Smith, C. (2007). 'Beliefs, knowledge and values held by inner-city youth about gardening, nutrition, and cooking', *Agriculture and Human Values*, Vol. *24*, no. 2, pp. 245-258.

Lawson, L. J. (2005). City Bountiful: A century of community gardening in America, University of California Press, Los Angeles.

Leone, L. A., Beth, D., Ickes, S. B., McGuire, K., Nelson, E., Smith, R. A., Tate, D. F. & Ammerman, A. S. (2012). 'Attitudes toward fruit and vegetable consumption and farmers' markets usage among low-income North Carolinians', *Journal of Hunger and Environmental Nutrition*, vol. 7, no. 1, pp. 64-76.

Lewis, L. B., Sloan, D. C., Nascimento, L. M., Diamant, A. L., Guinyard, J. J., Yancy, A. K. & Flynn, G. (2005). 'African Americans' access to healthy food options in South Los Angeles restaurants', *American Journal of Public Health*, vol. 95, no. 4, pp. 668-673.

Litt, J. S., Soobader, Mah. J., Turbin, M. S., Hale, J. W., Buchenau, M. & Marshall, J. A. (2011). 'The Influence of Social Involvement, Neighbourhood Aesthetics, and Community Garden Participation on Fruit and Vegetable Consumption', *American Journal of Public Health*, vol. 101, no. 8, pp. 1466-1473.

Loopstra, R. & Tarasuk, V. (2013). 'Perspectives on Community Gardens, Community Kitchens and the Good Food Box Program in a Community-based Sample of Low-income Families', *Canadian Journal of Public Health*, vol. 104, no. 1, pp. e55-e59.

Lyson, T. A. (2004). *Civic agriculture: Reconnecting, farm, food, and community*, Tufts University Press. Medford, MA.

McCormack, L. A., Laska, M. N., Larson, N. I. & Story, M. (2010). 'Review of the nutritional implications of farmers' markets and community gardens: A call for evaluation and research efforts', *Journal of American Dietetic Association*, vol. 110, no. 3, pp. 399-408.

McPhail, D., Chapman, G. E. & Beagan, B. I. (2013). 'The rural and the rotund? A critical examination of food deserts and rural adolescent obesity in the Canadian context', *Health and Place*, vol. 22, pp. 132-139.

Morland, K. W. S., Diez Roux, A. & Poole, C. (2002). 'Neighbourhood characteristics associated with location of food stores and food service places', *American Journal of Preventative Medicine*, vol. 22, pp. 23-29.

Morland, K. W. S., Wing, S. & Diez Roux, A. (2002). 'The contextual effect of the local food environment on residents' diets: The Atherosclerosis Risk in Community Study', *American Journal of Public Health*, vol. 92, pp. 1761-1767.

Morris, J. L., Neustadter, A. & Zidenberg-Cherrr, S. (2001). 'First grade gardeners more likely to taste vegetables', *California Agriculture*, vol. 55, pp. 43-46.

Morris, J. L., Koumjian, K. L., Briggs, M. & Zidenberg-Cherrr, S. (2001). 'Nutrition to grow on: A garden enhanced nutrition education curriculum for upper-elementary schoolchildren. *Journal of Nutrition and Education Behaviour*, vol. 34, pp. 175-176.

Morris, J. L. & Zidenberg-Cherrr, S. (2002). 'Garden-enhanced nutrition curriculum improves fourth-grade school children's knowledge of nutrition and preferences for some vegetables', *Journal of American Dietetic Association*, vol. 102, pp. 91-93.

Nelson, T. (1996). 'Closing the nutrient loop: using urban agriculture to increase food supply and reduce waste', *World Watch*, 9, Nov/Dec, 1996, (pp. 10-17).

Patel, I. C. (1991). 'Gardening's socio-economic impact', *Journal of Extension*, 29. Available from: http://www.joe.org/joe/1991winter/a1html (October 15, 2017).

Pasquani, M. & Young, E. M. (2009). 'Preface', in CM. Shackleton, MW Pasquani & AW Drescher, (eds.) *African Indigenous Vegetables in Urban Agriculture*, pp. xxi-xxvi, Earthscan, London.

Payne, G. H., Wethington, H., Olsho, L., Jernigan, J., Farris, R. & Walker, D. K. (2013). 'Implementing a farmers' market incentive program: Perspectives on the New York City health bucks program', *Preventing Chronic Diseases*, vol. 10, pp. 120-285.

Pennsylvania Horticultural Society (PHS). (2009). '*Community gardens of the 21st Century. Growing for the future. Strategy for a green city*', Pennsylvania Horticultural Society.

Riboli, E. & Norat, T. (2003). 'Epidemiologic evidence of the protective effect of fruit and vegetables on cancer risk, *American Journal of Clinical Nutrition*, vol. 78, pp. 559S-569S.

Sen, A. (1981). 'The great Bengal famine', in A. Sen & D. Omnibus (eds.), Poverty and famines, hunger and public action; *India: Economic development and social opportunity*, pp. 52-85, Oxford University Press, Oxford.

Tong, D., Ren, F. & Mack, J. (2011). 'Locating farmers' markets with an incorporation of spatio-temporal variation', *Socio-Economic Planning Sciences*, vol. *46*, pp. 149-156.

United States Department of Agriculture (USDA). (2009). 'Access to affordable and nutritious food: Measuring and understanding food deserts and their consequences. A report to Congress', *USDA*, Washington, DC.

Wakefield, S., Yeudall, F., Taran, C., Reynolds, J. & Skinner, A. (2007). 'Growing urban health: Community gardening in south-east Toronto', *Health Promotion International*, vol. *22*, no. 2, pp. 92-101.

Wang, H., Feng, Q. & Swallow, B. (2014). 'Can community gardens and farmers' markets relieve food desert problems? A study of Edmonton, Canada', *Applied Geography*, vol. *55*, pp. 127-137.

Zenk, S. N., Shultz, A. J., Israel, B. A., James, S. A., Bao, S. & Wilson, M. L. (2006). 'Fruit and vegetable access differs by community and racial composition and economic position in Detroit, Michigan', *Ethnic Dis*, vol. 16, pp. 275-280.

Zenk, S. N., Shultz, A. J. & Hollis-Neely, T. (2005). 'Fruit and vegetable intake in African Americans: Income and store characteristics', *American Journal of Preventative Medicine*, vol. *29*, pp. 1-9.

In: Agriculture, Food, and Food Security
Editor: Clinton Lloyd Beckford
ISBN: 978-1-53613-483-4
© 2018 Nova Science Publishers, Inc.

Chapter 7

ASSESSING VULNERABILITY TO CLIMATE CHANGE ACROSS AGROECOLOGICAL ZONES IN JAMAICA: INSIGHTS FOR COMMUNITY-BASED ADAPTATION STRATEGIES

Donovan Campbell[*]
Department of Geography and Geology
University of the West Indies, Mona, Kingston, Jamaica

1. INTRODUCTION

1.1. Background

Despite its small size, Jamaica has a number of agro-ecological zones where farming conditions differ, and a mosaic of agricultural systems dependent on local precipitation, soil, and topographic conditions. Given the

---

[*] Corresponding author email: donovan.campbell@uwimona.edu.jm.

different agro-ecological and socio-economic conditions of farming communities across the island, the likely effects of climate change on local agriculture will vary significantly from place to place, with significantly different impacts on livelihood security. Incorporating micro-scale variations can enhance the effectiveness of climate resilient programmes in farming communities. In this chapter, a livelihood vulnerability assessment approach is used to identify entry points for building climate resilience in 12 farming communities across nine agro-ecological zones in Jamaica. The vulnerability assessment is based on a survey of farming households (n=618) and focus group discussions in each community to assess levels of livelihood exposure to climate variability and change, climate impacts and adaptive capacity. The results exemplify the place and context-specific nature of vulnerability and demonstrate how this information can be used to shape current and future climate resilient strategies by aligning community needs with adaptation priorities.

1.2. Climate Change and Caribbean SIDS

The vulnerability of Caribbean Small Island Developing States (SIDS) to climate change is well established (Scandurra et al. 2018; Beckford & Rhiney 2016; Kelman 2014; Barker 2012; Pulwarty et al. 2010). Climate change in the Caribbean region is expected to result in warming of between 1.4 and 3.2 degrees by the end of the 21st Century (Climate Studies Group 2013). In the last decade or so, increased climate variability has complicated the livelihoods of many farmers across the Caribbean, and interrupted pathways to economic development. The vulnerability of agriculture-based livelihoods to climate change is likely to vary significantly across the region.

These livelihoods are often paradoxically characterized as both vulnerable and resilient to these changes (Osbahr et al. 2010; Ellis 1988). This paradox recognizes that livelihood vulnerability and resilience have both internal and external components and are driven by interrelated and often complex socio-economic, political, and environmental issues.

The role of agriculture as a vehicle for economic expansion in the Caribbean, has been declining in recent times. On average, agriculture contributes 10 percent to the GDP of Caribbean countries. However, the economic contributions of the sector vary significantly across the region from as low as 0.5 percent of GDP in Trinidad and Tobago, to as high as 20 percent in Guyana (FAO 2016). Despite its low (and declining) contributions to GDP, agriculture remains a significant source of employment, especially in the Greater Antilles (Cuba, Dominican Republic, Haiti, and Jamaica). For example, in the Dominican Republic, agriculture declined from 21 percent of GDP to five percent between 2000 and 2016 but employed 16 percent of the population. In Jamaica, agriculture contributes 7.3 percent to GDP but employs 19 percent of the population (Planning Institute of Jamaica 2017). In Haiti, the poorest country in the western hemisphere, approximately 20 percent of the population work in agriculture, which accounts for 22 percent of GDP.

Climate models suggest that the northern Caribbean will have a wetter November to January period, while the southern Caribbean is projected to be drier (Taylor, James & Stephenson 2017). The general scientific consensus is that the Caribbean region is experiencing a more variable climate of stronger dry season droughts (Herrera & Ault 2017) and more-stormy wet season condition (Taylor, James & Stephenson 2017, Climate Studies Group 2012). The impacts of climate change on agriculture production, food, and livelihood security in the developing world, is highly uncertain and requires systematic vulnerability assessments.

Susceptibility to natural hazards is not uniform throughout the region. The Greater Antilles (Cuba, Haiti, and Jamaica) have been identified as the most disaster-prone group in the Caribbean and along with Pacific islands with unstable economies and weak political and institutional development, are heavy losers to repeated natural shocks (Pelling & Uitto 2001). This is supported by Spence et al. (2005) who found that the northern Caribbean including Jamaica, Cuba, Hispaniola, Puerto Rico, The Bahamas, Turks and Caicos Islands, and the Cayman Islands, has high inter-annual variability of hurricane occurrence with a mean strike rate of 1 per year; while the southern

Caribbean experiences a much lower strike rate of 0.4 hurricane strike per year.

Hurricanes are the most prevalent meteorological hazards that occur in the Caribbean (Poncelet 1997; Pielke et al. 2003). The global hurricane belt includes all tropical oceans between latitudes 40 degrees south to 40 degrees north except the southern Atlantic. However, some researchers suggest that "all portions of Latin America (including Central America and South America) south of 10 °N had a less than 1percent chance of a hurricane strike per year. The annual likelihood of hurricane activity increased farther from the equator to a maximum of >20 percent northeast of The Bahamas" (Pielke et al. 2003 p.102). The 2017 hurricane season highlighted the vulnerability of the Caribbean when several islands suffered two Category 5 hurricanes (Irma and Maria).

1.2. Local Trends and Implications

In Jamaica, there is increasing concern over the impacts of climate change and increased climate variability on agricultural production and food security. The Near-Term Climate Scenarios (2014) report for Jamaica, highlights that observed sea level rise from 1950 to 2000 indicate that the rise in the Caribbean is near the global mean. Warming trends were found to be consistent with a warming pattern across the global (Caribbean Community Climate Change Center [CCCCC] 2016). Data from the airport stations indicate historical warming of 0.20 – 0.31 °C per decade, with the greatest warming occurring between June and August. This trend is validated by a recent analysis of sea surface temperatures around Jamaica where a statistically significant annual increase of +0.15C/decade from 1980 to 2016 was observed (CCCCC 2016). The frequency of 'hot' days and nights is expected to increase, reaching 30-98% of days annually by the 2090s while 'Cold' days and nights are projected to diminish in frequency, occurring on a maximum of 2% of days/nights by the 2080s (Climate Studies Group 2012). As far as droughts are concerned, there has been a tendency towards longer and more intense droughts in Jamaica since the early 1990s (Gamble et al. 2017; Gamble et al. 2010).

While these trends are important to understand, climate models and projections are only parts of the climate change picture which is characterized by unique and complex interactions between the biophysical environment and human systems. Issues related to adaptation, vulnerability, and resilience, have become dominant themes in the climate change discourse. Incorporating the human dimensions of climate change presents a unique opportunity to assess the success of vulnerability reduction and climate change adaptation programmes. Because vulnerable groups face inherent adaptation challenges, and are the ones who are expected to need climate change adaptation the most, it is important that adaptation priorities are set by them.

The contemporary vulnerabilities of agriculture and rural communities in Jamaica are both economic and environmental in nature. The interaction of these factors can heighten vulnerabilities at both the national and community level of analysis. The Fifth International Panel on Climate Change (IPCC) Report reaffirms the need to assess the vulnerability of local communities and their abilities to adapt and cope by investigating and incorporating local knowledge and identifying local good practices (IPCC 2014). A critical overview of the status of agriculture in Jamaica and the sector's vulnerability to climate variability and change, shows that climate change is a cause for serious concern. Agriculture is still the principal source of income for rural communities across the island. Therefore, the effects of climate change will significantly impact rural livelihoods that depend on agriculture-related ecosystem services.

To effectively assess vulnerability and devise appropriate adaptation strategies, a 'bottom-up'/grassroots research approach is essential. It is important to recognize that answers to adaptation questions are not always present at the community level and may be situated in broader institutional processes and structures. It is therefore important to adopt a pragmatic research approach when assessing climate change vulnerability. While community-based vulnerability assessments have become more popular in Jamaica, and has illuminated important micro-scale issues, integrated macro-scale analyses are generally lacking. Agriculture is one of the most vulnerable sectors to climate change. The current vulnerability assessments through traditional fragmented sectoral methods are insufficient to capture the effects on complex agricultural systems. Therefore, these traditional methods need to be replaced with more integrated approaches. The objective of this study is to propose a holistic vulnerability assessment method for agricultural systems in Jamaica.

2. Methodology

This research represents the most extensive agriculture-based vulnerability assessment to be conducted in Jamaica (to date), and is anticipated to provide valuable baseline information to support community-based climate resilient intervention. In this research, a livelihood vulnerability assessment framework is used to provide estimates of the degree of vulnerability that exists in farming communities located in different agro-ecological zones across Jamaica. The complex and dynamic nature of small farming dictates the use of both qualitative and quantitative techniques to effectively assess livelihood vulnerability to climate change. The methodology for this study was guided by the principles and philosophy of mixed methods research. Mixed research is the third major research paradigm which represents a pragmatic alternative to qualitative and quantitative designs. Mixed research is grounded in the compatibility thesis and the philosophy of pragmatism (Bazeley 2004). The compatibility thesis hinges on the idea that qualitative and quantitative methods are compatible and can be used in a single study.

Understanding the complexity of livelihood vulnerability to climate change in agricultural communities requires an assessment of exposure, sensitivity (potential impact) and adaptive capacity. Here, these three main components are used to develop a composite index to profile livelihood vulnerability in the study communities. A total of seven sub-components and 30 indicators were used to represent the three main components (see Table 7.2.1). For the purpose of analysis, it was deemed necessary to combine exposure and sensitivity as a single component because of their close association. Smit and Wandel (2006 p.286) observe that sensitivity and exposure are "almost inseparable properties of a system and are dependent on the interaction between the characteristics of the system and on the attributes of the climate stimulus".

The conceptualization of adaptive capacity is guided by (and modified from) the approach and indicators used by Hahn et al. (2009). Many community residents already employ adaptive actions for dealing with climate change. Thus, the existing adaptive capacity, as well as the knowledge, attitudes and perceptions of householders regarding climate change were also captured. An important feature of the research is the development of a standard framework and methodology for assessing and comparing livelihood vulnerability at the household and community levels. The sustainable livelihoods approach, forms the conceptual base for developing indicators based on household assets. Methods used allow for the integration of local knowledge, especially individual and community perceptions of climate change and local coping strategies, using both quantitative and qualitative approaches to data collection and data analysis.

2.1. Research Approach

The methodological approach used for the livelihood vulnerability assessment is guided by the principles of the sustainable livelihoods approach (SLA). The sustainable livelihoods approach uses both qualitative and quantitative techniques to develop a holistic framework to improve knowledge and understanding of the livelihoods of poor people. The approach is rooted in the principles of participatory learning and seeks to capture vulnerabilities of agricultural households within the context of climate variability and change. The twelve communities were selected under the Jamaica Rural Economy and Ecosystems Adapting to Climate Change (Ja REEACH) based on priority interventions identified in consultation with national stakeholders. The objective of the Ja REEACH programme is to promote the protection of rural lives, livelihoods, and ecosystems, through interventions that increase and strengthen climate change adaptation.

Table 7.2.1. Vulnerability indicators

	Main components	Sub-components	Indicators
Vulnerability	Adaptive Capacity	Socio-demographic factors	Dependency ratio
			Education/training
			Female-headed households
		Livelihood strategies and flexibility	Sources of Income
			Livelihood Diversification
			Dependence on livelihood activity
		Social network/organization	Support within livelihood activity/or community
			Support outside of community
			Participation in community group or organization
			Government support services (access)
			Community support services
		Assets	Access to livelihood resource
			Ownership of house
			Ownership of land
			Loan/credit
			Insurance
			Saving
			Livelihood assets
		Knowledge, awareness & learning	Disaster risk communication
			Information on market price and condition
			Use of technology
			Disaster mitigation strategies
	Potential Impact	Biophysical	Production failures
			Condition of natural resource
			Climate change & variability
		Socio-economic	Input cost and access
			Livelihood conditions
			Health conditions
			Food access stability
			Water access

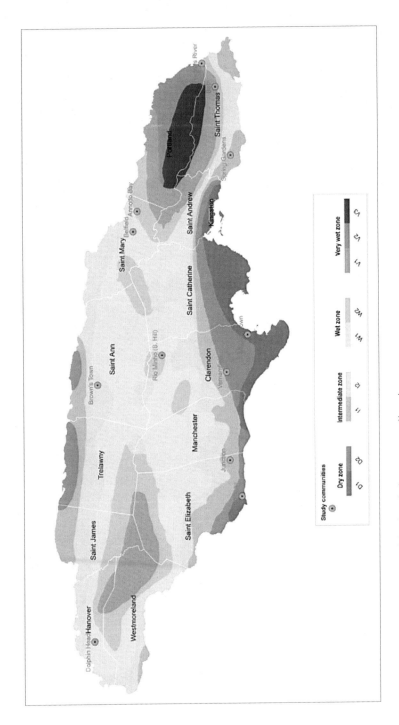

Figure 7.1. Location of communities in relation to agro-climatic zones.

Table 7.2.2. Community sample size, agroecological zone and parish

Parish	Community	Sample size	Agro-climatic zone
Hanover	Dolphin Head	68	W1
Clarendon	Rio Minho	55	W2
St. Ann	Brown's Town	39	I2
St. Elizabeth	Flagaman	50	D1
St. Thomas	Bath	40	V2
Westmoreland	Annotto Bay	63	W2
St. Thomas	Spring Garden	50	I1
Portland	Hectors River	40	V2
St. Mary	Bellfield	47	W1
Clarendon	Mitchell Town	63	D1
St. Elizabeth	Junction	71	I1
Clarendon	Vernamfield	32	D2

In the selection of the communities, consideration was given to the specific Ja REEACH programme being implemented, targeted parish and the agro-climatic setting. Figure 7.1 shows the location of the selected communities and their agro-climatic zone.

The overall methodology was divided into four main phases- preparatory, exploratory, assessment, and data analysis. The preparatory phase was aimed at increasing awareness and establishing partnerships with community members and other key stakeholders operating at the local level. Ongoing Ja REEACH intervention program in the selected communities meant that the communities were already sensitized about the programme. The exploratory phase consisted of a review of available documents and data from key agencies such as the Social Development Commission (SDC), Office of Disaster Preparedness and Emergency Management (ODPEM), Statistical Institute of Jamaica (STATIN), the Ministry of Agriculture, the Planning Institute of Jamaica and the National Environment Planning Agency (NEPA). Because the quality and value of any vulnerability assessment is dependent on the indicators selected, this phase also involved validating and fine-tuning the indicators with stakeholders from these organizations.

In the assessment phase, both qualitative and quantitative assessments were conducted. Questionnaire surveys were the main method used to collect quantitative, data. A total of seven sub-components and 30 indicators were used to represent adaptive capacity and the potential impacts of climate change. In each community, 50 percent of the districts were selected randomly and a census approach was taken to administer the questionnaires within the community boundaries (as defined by the farmers) (see Table 7.2.2). The household survey instrument was designed to capture information on household characteristics, farming activities, livelihood strategies resource management strategies, natural hazards and climate change, mitigation and adaptation practices.

The design and structure of the questionnaire also facilitated the acquisition of qualitative data (for example, farmers were asked about their attitudes and perceptions to the pertinent hazards) primarily with the use of open-ended questions. However, the bulk of the qualitative data were collected via focus group discussion, key informant interviews, and participant observation. The selection of participatory tools was guided by the type of vulnerability information they are designed to capture. Seasonal calendars were used to capture exposure, hazard ranking and impact to assess sensitivity, and vulnerability and capacity matrix along with coping and adaptation strategies assessment to determine adaptive capacity at the community level. A deliberate attempt was made to ensure group heterogeneity and balance between male and female participants. The results presented here are primarily based on the quantitative assessment.

In terms of the data analysis phase, the quantitative data were managed and explored in the Statistical Package for the Social Sciences (SPSS). The components of vulnerability were aggregated and deferential characteristics (e.g., gender and age) explored. The analysis of indicators involved a very careful process combining baseline survey data with variables informed by the literature. However, before the variables were aggregated into a composite index, certain statistical procedures were undertaken. Such procedures involved techniques of variable computation, transformation and standardization. Once all the variables were sorted, the sub-component variables were averaged to calculate the value for each major component.

By averaging the major components to compute livelihood vulnerability index for each community, the major components were combined to reflect exposure-sensitivity and adaptive capacity. The qualitative data analysis involved documenting interpretations, seasonal calendars, hazard ranking, vulnerability and capacity matrix and existing coping and adaptation strategies assessment.

3. RESULTS AND DISCUSSION

A livelihood vulnerability index was developed for the four communities using data collected from various aspects of the field research. Data related to livelihood strategies and flexibility, social network and organizations, assets, knowledge, awareness and learning, as well as socio-demographic, biophysical and socio-economic factors were aggregated using a composite index, and differential vulnerabilities were compared. Table 3.1 provides a summary of the community index values.

Table 7.3.1. Vulnerability component scores at the community level

Community	Vulnerability score	Rank (least to most)
Hectors River	0.12	12
Dolphin Head	0.38	11
Brown's Town	0.38	10
Vernamfield	0.45	9
Annotto Bay	0.49	8
Bath	0.50	7
Rio Minho	0.50	6
Mitchell Town	0.59	5
Bellfield	0.61	4
Junction	0.66	3
Spring Garden	0.81	2
Flagaman	0.85	1

The analysis of vulnerability data for the 12 communities revealed Flagaman as the most vulnerable. Flagaman has the highest exposure-sensitivity score and the lowest adaptive capacity. Junction and Belfield exhibit relatively high vulnerability characteristics. Junction is ranked as the third most vulnerable community in the sample. Belfield has the highest adaptive capacity and the third lowest exposure in the overall sample. When combined with a relatively high sensitivity (4th highest overall), Belfield emerged as the 4th most vulnerable community. Mitchell Town, Rio Minho and Bath are classed as moderately vulnerable. Mitchell Town has a relatively high adaptive capacity, a moderate level of sensitivity, and very high level of exposure.

With index scores of 0.66 and 0.61 respectively, Junction and Belfield exhibit high vulnerability relative to the other communities. Despite having the 5th lowest adaptive capacity, 4th highest exposure, and 5th highest sensitivity, Junction is ranked as the third most vulnerable community in the sample. Belfield has the highest adaptive capacity and the third lowest exposure in the overall sample. When combined with a relatively high sensitivity (4th highest overall), Belfield emerged as the 4th most vulnerable community in the overall sample. Mitchell Town, Rio Minho and Bath are classed as moderately vulnerable. Mitchell Town has a relatively high adaptive capacity, a moderate level of sensitivity, and a very high level of exposure.

The Rio Minho community exhibited the second highest level of adaptive capacity, 4th highest exposure, and the highest degree of sensitivity in the overall sample. When combined, the Rio Minho communities are ranked 6th in the overall sample. Bath is ranked as the 7th most vulnerable community. Annotto Bay and Vernamfield show relatively low vulnerability, while Brown's Town, Dolphin Head and Hector's River, are classed as very low. The livelihood vulnerability values were analyzed in the Geographic Information Systems (GIS) computer software ArcGIS (version 9.0) to produce community-level vulnerability maps as well as maps of the IPCC's framework components of exposure, adaptive capacity and sensitivity (Figure x-x).

Figure 7.2. Patterns of adaptive capacity, exposure-sensitivity and vulnerability.

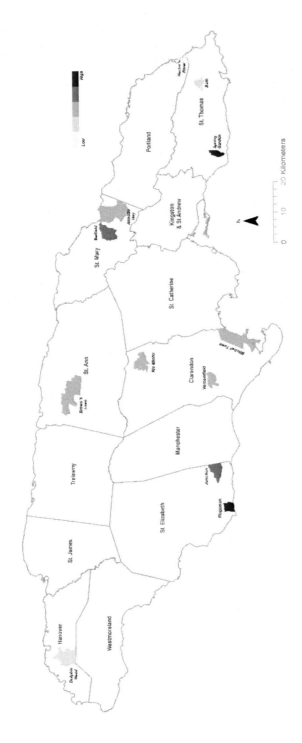

Figure 7.3. Spatial pattern of community vulnerability.

Figure 7.4. Spatial pattern of community exposure-sensitivity.

Figure 7.5. Spatial pattern of community adaptive capacity.

3.1. Determinants of Vulnerability

3.1.1. Socio-Demographic Factors

The socio-demographic indicator of livelihood vulnerability is a composite of variables representing the percentage of female-headed households; level of education and training dependency ratio; and, the percentage of households dependent on one livelihood activity. The aggregation of these variables revealed Brown's Town and Spring Garden as the most vulnerable, while Hector's River was deemed the strongest in this category. Such findings may be further explained by analyzing the individual community profiles, highlighting the underlying factors contributing to these outcomes. See Figures 7.6-7.11.

In the Brown's Town area, high socio-demographic vulnerability is partly explained by the social structure of the community. A relatively high proportion of community members (84%), dependent on only one livelihood activity, which has resulted in its ranking as the second least economically diversified community in the sample. Additionally, Brown's Town has the highest proportion of respondents (90%) with no formal training related to their livelihood activity, and the second highest dependency ratio in the overall sample. This implies that a relatively high percentage of the population is dependent on the working-age cohort, resulting in a greater demand from individuals who make a living from ecosystem services. In the Caribbean, single-parent female-headed households are considered to be more vulnerable than other family arrangements. In this case, Brown's Town has the fourth highest percentage of female-headed households.

Spring Garden exhibits the highest dependency ratio in the overall sample. With 70 percent of its respondents reporting that their household is dependent on one livelihood activity, Spring Garden is the third least economically diversified community in the sample. This lack of economic livelihood diversity, further exacerbates pressure on the working class and ecosystem services. Conversely, Hector's River exhibits the lowest socio-demographic vulnerability in the overall sample. Hector's River has the lowest dependency ratio meaning that it has a relatively small population that depends on the working population. As a result, individuals with

Assessing Vulnerability to Climate Change ...

livelihoods associated with ecosystem services face less pressure from its dependent population, and with 49% of the population relying on a sole livelihood activity, Hector's river is classified as the third most economically diverse community in the sample.

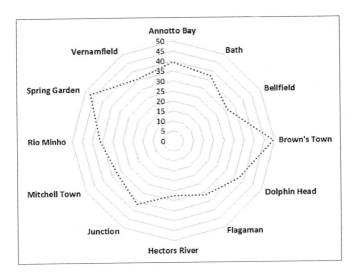

Figure 7.6. Socio-demographics.

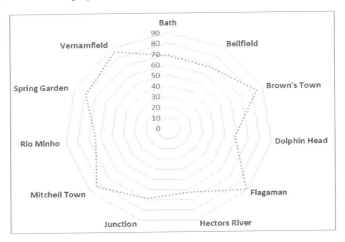

Figure 7.7. Social network & organization.

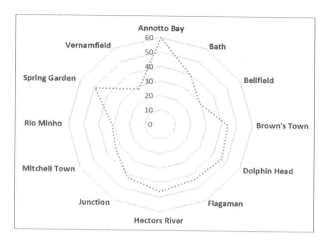

Figure 7.8. Knowledge & Awareness.

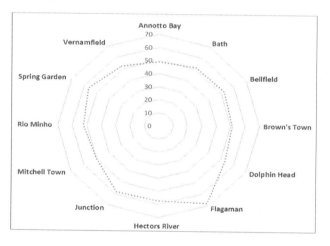

Figure 7.9. Asset-base.

3.1.2. The Role of Social Network and Organization

The indicator representing social network and organization, is a combination of factors representing livelihood and community support; participation in local groups and organizations; and access to governmental and external support services. Based on this indicator, Flagaman and Vernamfield, emerged as the two weakest communities in the sample, while Dolphin Head exhibited the strongest characteristics. Flagaman's ranking is mainly influenced by the low level of livelihood support from inside and outside of the community. In both cases, Flagaman has the lowest and the third lowest respectively. Level of participation in community groups is also very low, and most respondents indicated that they receive minimal support from the government. Only 8% of the respondents indicated that they have ever received support from the government with their livelihood activities.

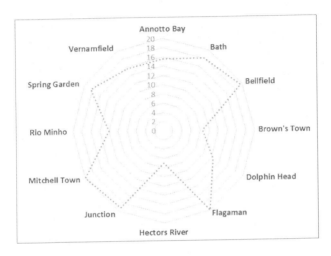

Figure 7.10. Biophysical determinants.

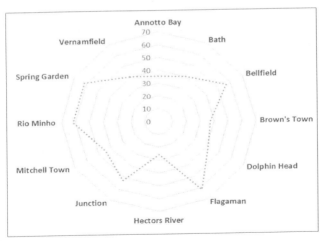

Figure 7.1. Socio-economic drivers.

Vernamfield's low ranking on the social network and organization indicator, is partly attributed to the community possessing the lowest level of support with livelihood activities, and the second lowest percentage of households with support outside the community. The low level of participation in community groups or organizations is also a major contributing factor with most of the respondents indicating that they prefer to operate individually. Additionally, only 32% of the respondents indicated that they received support/assistance from the government in the past 5 years. Conversely, the Dolphin Head area exhibits the strongest social network, and the highest percentage of individuals participating in community groups and organizations. The community also has the third highest level of support from outside the community, and the second highest level of support with livelihood activities.

3.1.3. Knowledge and Awareness

Knowledge and awareness are important influencing factors in decision-making. This is an important element in a community's or individual's capability to reduce exposure to climate stressors and overall vulnerability. To gauge levels of awareness and knowledge, the study assessed the extent to which communities are aware of climate change and the impacts associated with it; if they practice disaster mitigation; and if they access or receive information on disaster risk conditions. The aggregation of indicator variables, reveals that Annotto Bay and Spring Garden have the lowest level of knowledge and awareness while Vernamfield has the highest.

Annotto Bay emerged as the community with the second lowest level of climate change awareness in the sample with 35% of the respondents indicating that they were unaware of the potential impacts of climate change. Despite the fact that 21% of the respondents indicated that they do not receive disaster risk information, 93% of them indicated that they practice some form of disaster mitigation. This is coupled with a general scarcity of information regarding market conditions with 90% of the respondents indicating that they do not receive information on agricultural market prices and conditions.

Spring Garden has the lowest level of climate change awareness in the sample. Specifically, 44% of the respondents indicated that they were unaware of the potential impacts of climate change. In the overall sample, Spring Garden also had the highest percentage of respondents who reported that they did not receive disaster risk information. Not only is there a lack of information regarding disaster risk, but more than half of the respondents (52%) indicated that they do not receive information on market prices, thus highlighting an overall information deficiency in this community.

With only 2% of the respondents indicating that they are unaware of the potential impacts of climate change, Vernamfield ranks highest in the knowledge and awareness categories. Vernamfield is the second most climate change aware community. Access to disaster risk information at the community level is relatively high with 96% of the respondents indicating that they receive some disaster risk information. On the other hand, market information is relatively low with 81 percent of the respondents indicating that they do not receive information on market prices and condition. Overall, a lack of information regarding market conditions, was experienced across all three communities with relatively high percentages of individuals being unaware of potential climate change impacts on farming livelihoods in these vulnerable communities.

3.1.4 Asset Base

The assets-based indicator of livelihood vulnerability, is a composite of variables representing ownership of physical assets (house and land ownership), and financial assets (loan access insurance, savings, remittances, and livestock). In addition to having the highest level of knowledge and awareness, Annotto Bay has the lowest level of asset vulnerability. This is partly attributed to the percentage of respondents (65%) indicating that they practice some form of financial saving. Conversely, the community has the second lowest percentage (77%) of respondents indicating that they have never received any loan or credit in the past and that they are unaware of mechanisms to do so.

Flagaman is the most vulnerable in this category and has the highest number of respondents with limited access to resources necessary to effectively carry out their livelihood activities. Having the second highest percentage (97%) of respondents indicating that they have never received a loan or credit in the past, further highlights the community's limited access to financial resources. Junction emerged as the second most vulnerable community in this category. This is partly explained by the fact that half of the respondents indicated that they found it difficult to access livelihood resources. Additionally, Junction has the highest percentage (98%) of respondents indicating that they have never received any loan or credit in the past and 92% lacking insurance of their major assets. A cross-cutting driver of vulnerability in the three communities is limited access to financial resources and safety-net such as loans and insurance.

3.1.5. Biophysical Exposure

Biophysical exposure, is measured as the degree of stress and impact of climate hazards on livelihood activities in the selected communities. Variables used to determine exposure to hazard such as hurricanes, floods, drought, rainfall uncertainty, and pests and diseases, are aggregated to form a composite indicator. The analysis revealed Mitchell Town and Flagaman as the most vulnerable and Hector's River as the least vulnerable. Flagaman's high biophysical exposure is attributed to the combined effects of drought and rainfall uncertainty on agricultural production in the community. Eight-two percent of the respondents indicated that they had been badly affected by drought in the last 10 years, and 61% highlighted rainfall uncertainty as a major hindrance to their productivity. In the past 4 years, plant pests have exacerbated the environmental challenges in the area, with the beet armyworm (*Spodoptera exigua*) being particularly problematic.

Mitchell Town is one of the most flood-prone communities in Jamaica, but the impact on the livelihoods of community members in recent times has been particularly severe. With 89% of respondents indicating that they have been badly affected by hurricanes (and extreme rainfall events) and ranked as the highest level of exposure in the sample, the community is highly exposed to multiple hazards. Drought and changes in the timing of dry season onset have been especially problematic for the community. In

Hectors River, community livelihood activities are highly dependent on ecosystem services. Drought and rainfall uncertainty are the two most problematic hazards affecting the livelihoods with 68% of the respondents having been badly affected by drought, and 50% by rainfall uncertainty. Flooding was not seen as a major concern as Hectors River has the lowest number of respondents affected by flooding. Only 3% of the community indicated having been badly affected by flooding.

3.1.6. Socio-Economic Exposure-Sensitivity

Sensitivity to socio-economic conditions is a combination of access to livelihood resources, access to adequate food, access to water, and prevalence of chronic health problems. The results indicate that Rio Minho and Spring Gardens have the highest levels of exposure sensitivity to socio-economic conditions, while Hectors River has the lowest. Market price uncertainty increases the likelihood of production failure for residents of the Rio Minho agroecological zone, especially farmers who are highly dependent on the declining sugar, and cocoa industries. Coupled with this high risk of market failure, are poor farm roads that inhibit increased agricultural production; lack of sufficient food to satisfy household needs, and health problems reported by 35% (mostly among the older farmers) of the respondents. Spring Gardens face similar challenges to Rio Minho with a relatively high proportion (62%) of respondents indicating that they face challenges obtaining enough food to satisfy the needs of their families.

Hector's River, on the other hand, has the lowest level of socio-economic exposure and sensitivity. This is attributed to the community having the lowest percentage of respondents reporting chronic health problems at the household level and the second highest level of access to clean water. However, these positive conditions are being threatened by the displacement of small-scale hillside food farmers by sugar cane cultivation. This has resulted in many small-scale food farmers finding it difficult to access sufficient food to feed their families or households. In addition to these challenges, the lack of reliable marketing facilities for domestic food produce, has relegated farmers to a reliance on market purveyors as the main outlet for their produce. These local food marketers known as 'higglers', can be unreliable which often causes problems for the farmers selling their produce and getting paid on time.

3.2. Lessons for Climate Resilient Strategies across the Communities

Successful community-based adaptation planning involves the development of locally specific (grass-root) climate change solutions. It is about responding to local needs and capturing lessons that can be replicated - scaled up and out. The process often involves international agencies working in partnership with governments and targeted local communities to build resilience to various stressors and shocks. The results of this assessment can be used to support efforts to: (i) engage communities to drive adaptation responses; (ii) strengthen community-based disaster resilience; and (iii) promote climate-smart agriculture.

3.2.1. Engaging Communities to Drive Adaptation Responses

The level of readiness of communities to participate in climate resilient interventions is an important sustainability prerequisite and safeguard for achieving intended objectives. Therefore, it is important to understand the most effective ways to engage communities to ensure representation of their interests and values and to empower them to drive climate change adaptation responses.

Findings from the vulnerability assessment in Mitchell Town and Spring Garden in particular, can be useful to this effort. Both communities have relatively high dependency ratios compared to the overall sample. This implies a relatively high dependent population on the working-age cohort. In these communities, individuals with livelihoods associated with ecosystem services face greater pressure from its dependent population. Dependency ratios are important in vulnerability assessments, because they provide useful insights into the general future economic and social health of the target population. Because the dependency ratio is influenced by both micro and macro processes, downward management should be approached from a similar dualistic perspective. While community-based vulnerability assessments have become more popular in Jamaica, and have illuminated important micro-scale issues, integrated macro-scale analyses are limited. At the national level, socioeconomic policy should recognize a dependency ratio range and a threshold beyond which appropriate and urgent interventions are activated.

Declining agricultural income, is one of the major factors causing increased pressure on the working-age cohort in these two communities. The promotion of sustainable alternative livelihoods in the communities is one strategy that can be implemented to reduce the dependency ratio and improve the adaptive capacity of the farming population. Agricultural livelihood diversification is relatively high in both communities and as such, there is a relatively high level of dependence on ecosystem services. The importance of promoting alternative livelihoods outside of agriculture as an important strategy to address intrinsic characteristics of vulnerability at this scale, is highlighted. In general, asset and livelihood diversification remain important strategies for risk reduction and increase options in the face of livelihood threats. Expansion of the ongoing apiculture program in Mitchell Town is one successful alternative livelihood solution that can be replicated in other communities.

Most of the respondents in the two communities (Mitchell Town and Spring Garden) have attained formal education but possess limited formal training related to their livelihood activities. In these communities, local traditional knowledge remains the most important livelihood asset and the key to managing natural resources including the agricultural base. Implementing livelihoods skills training programs and sharing good practices can be useful for developing adaptation response.

Participation in groups and organizations and cooperative activities were found to be very low in both communities. The broader impact and success of adaptation programmes will ultimately depend on who within the community can participate in the process. In particular, gender emerged as an important consideration based on the statistically differentiated impacts of climate shocks on men and women. One of the major findings regarding gender is that there is a greater proportion of female involvement in community groups and organizations compared to males (relative to the total number in each category). This is significant in terms of promoting participatory action at the community level.

Despite this, female farmers in these communities continue to be one of the most underutilized resources in development groups across the island. The success of the adaptation planning in general will be dependent on success tapping into women's unique abilities for building capacity and driving adaptation processes. This is aligned with a need to support institutional capacity development among farming groups (e.g., Jamaica Agricultural Society [JAS] and Producer Marketing Organizations [PMO]), an area where it is becoming increasingly apparent, that women's leadership hold the key to success.

3.2.2. Strengthening Community-Based Disaster Resilience

One of Jamaica's national policy objectives is to reduce vulnerability to disasters by improving disaster-risk-reduction strategies and the capacity of people to protect themselves, their assets, and their livelihoods. In agriculture, the aim is to build resilience to agrometeorological hazards through climate-smart solutions. Key lessons for strengthening community-based disaster resilience emerged from the assessments in Flagaman and Annotto Bay. The two communities are exposed to hazards at the extreme ends of the climate spectrum. Flagaman is located in a rain shadow area and, as such, experiences a dry micro-climate and recurring drought conditions. On the other hand, Annotto Bay is low-lying and is particularly susceptible to coastal and ravine flooding.

Despite these two extremes, the problem of rainfall uncertainty is a cross-cutting issue emerging from the assessment. In both Annotto Bay and Flagaman, rainfall uncertainty has presented various challenges to agricultural production. One recommendation here, is to improve access to agrometeorological information among farmers in order to minimize the impact of increased climate variability. Downscaling climate information, creating weather forecasts, and effectively communicating them to agricultural producers, is an important step in that direction. The agricultural meteorology programme at Caribbean Institute for Meteorology and

Hydrology (CIMH), hosts a drought and precipitation monitoring platform that can be utilized for this purpose. Understanding the local climatic rhythm by integrating scientific data into their local knowledge schema, can help farmers to manage resources more efficiently, reduce the impact of weather uncertainty, and capture market opportunities.

3.2.3. Promoting Climate-Smart Agriculture

Climate-smart agriculture (CSA) is about enhancing food security in a context of changing climate, and evidence from other parts of the world shows that the uptake of CSA practices is less likely in communities that face food insecurity challenges. Uptake of CSA practices is also linked to the institutional capacity of communities and is built on the pillars of sustainable development and promotes the integration of agricultural productivity/incomes and climate change resilience building. Lessons from six communities in the sample (Dolphin Head, Rio Minho, Hectors River, Bellfield; Junction and Vernamfield), provide useful insights for promoting climate-smart agriculture in Jamaica.

The findings from the assessment revealed important lessons that are directly related to levels of climate change and awareness as well as the socio-demographic factors and general wellbeing of the study communities. Across the sample, most of the respondents indicated that they often face challenges accessing the food they need on a daily basis. Participation in community groups and organizations was found to be very low among the sampled respondents. Additionally, most of the respondents lack formal training related to their livelihood activities. In these circumstances, CSA that promotes low-tech interventions should be prioritized over high-tech ones that require great technical expertise, and significant expenditure. Additionally, the context-specific nature of the vulnerability assessment enables the effective targeting of marginalized groups of farmers at the community level- an important consideration for effective CSA initiatives. The idea of building resilience and empowering marginalized groups is embedded in the CSA concept. The broader impact and CSA will ultimately depend on how broadly the farming community can participate in intervention strategies.

CONCLUSION

It is now widely accepted that vulnerability is a product of complex and context-specific human-environment interactions within and between sectors. This research has demonstrated that agricultural livelihoods in the targeted communities are exposed to risks of a different magnitude that they have to cope with on a daily basis in order to survive. Some individuals lack the capacity to meet the demand and expectations of formal institutional arrangements. Some 'smaller risks' are embedded in the social fabric of these communities and, in most cases tend to have the greatest impact on vulnerability. Strengthening social capital within these communities is an important step towards capacity development within the sector as a whole. The analysis presented, is an important step towards understanding the nature of vulnerability in rural farming- based communities across Jamaica. Assessments such as these could be used as a tool by policymakers to evaluate the success of climate change vulnerability reduction programs at the community level.

The ramifications of climate change impacts on agricultural livelihoods and communities are situated within a much broader context of struggling local and regional institutions. National and regional institutions continue to struggle with negative changes in social structures and international market conditions, which also fundamentally influence the adaptive capacity to cope with environmental change. The socioeconomic challenges of increasing livelihood options in their communities are at the forefront of the agricultural producer's concerns. The results from the analysis of the livelihoods vulnerability index, have shown that households possess considerable amounts of physical, natural, and social capital but, that gaps exist where financial and human capital are concerned. Intervention programmes should therefore, seek to enhance the human and financial capital in these communities.

When studying or assessing vulnerability problems, the behavior and livelihood patterns of farmers do not necessarily conform to traditional ideas of rationality. The key livelihoods characteristics of agricultural households exemplify the high degree of variability that exists at the local level and thus, the importance of utilizing a pragmatic research approach geared to local needs. A synthesis of the findings from this study reaffirms the importance of social capital and local knowledge in understanding livelihood vulnerability to environmental change in rural communities. An intensification for the support of livelihood diversification strategies in rural development planning is highly recommended here.

REFERENCES

Barker, D 2012, 'Caribbean Agriculture in a Period of Global Change: Vulnerabilities and Opportunities', *Caribbean Studies*, https://doi.org/10.2307/41917603.

Bazeley, P 2004, *Issues in Mixing Qualitative and Quantitative Approaches to Research. Applying qualitative methods to marketing management research.* https://doi.org/10.1007/s10447-006-9022-5.

Beckford, CL & Rhiney, K 2016, *Globalization, Agriculture and Food in the Caribbean: Climate Change, Gender and Geography.* Palgrave Macmillan, https://doi.org/10.1057/978-1-137-53837-6.

Beckford, CL Rhiney, K 2016, 'Geographies of globalization, climate change and food and agriculture in the Caribbean', in CL Beckford and K Rhiney (eds.) *Globalization, Agriculture and Food in the Caribbean: Climate Change, Gender and Geography.* https://doi.org/ 10.1057/978-1-137-53837-6_1.

Beckford, CL, Rhiney, K 2016, 'Future of Food and Agriculture in the Caribbean in the Context of Climate Change and Globalization: Where Do We Go from Here?', in CL Beckford and K Rhiney (eds.) *Globalization, Agriculture and Food in the Caribbean: Climate Change, Gender and Geography.* Palgrave Macmillan https://doi.org/ 10.1057/978-1-137-53837-6_11.

Beckford, C, Barker, D 2007, 'The role and value of local knowledge in Jamaican agriculture: Adaptation and change in small-scale farming', *Geographical Journal*, https://doi.org/10.1111/j.1475-4959.2007.00238.x.

Caribbean Community Climate Change Centre 2016, *Vulnerability and Capacity Assessment of Jamaica's Fisheries Sector*. Belmopan, Belize.

Climate Studies Group Mona (CSGM) 2012, *State of the Jamaican Climate 2012: Information for Resilience Building*. Kingston, Jamaica.

Climate Studies Group Mona (CSGM) 2014, *Near-Term Climate Scenarios for Jamaica* (Technical Report). Kingston, Jamaica.

Gamble, D, Burrell, D, Popke, J, Curtis, S 2017, 'Contextual analysis of dynamic drought perception among small farmers in Jamaica', *Climate Research* 74, 109–120. https://doi.org/https://doi.org/10.3354/cr01490.

Gamble, DW, Campbell, D, Allen, TL, Barker, D, Curtis, S, McGregor, D, Popke, J 2010, 'Climate change, drought, and Jamaican agriculture: Local knowledge and the climate record', *Annals of the Association of American Geographers*, https://doi.org/10.1080/00045608.2010.49712 2 .

Ellis, F 1988, *Peasant Economics*, Wye Studies in Agricultural and Rural Development.

FAO 2016, *FAOSTAT*, Food and Agricultural Organization of the United Nations: Rome.

Hahn, MB, Riederer, AM, Foster, SO, 2009, 'The Livelihood Vulnerability Index: A Pragmatic Approach to Assessing Risks from Climate Variability and Change - A Case Study in Mozambique', *Global Environmental Change*, https://doi.org/10.1016/j.gloenvcha.2008. 11.002.

Herrera, D, Ault, T 2017 'Insights from a new high-resolution drought Atlas for the Caribbean spanning 1950-2016', *Journal of Climate*, https://doi.org/10.1175/JCLI-D-16-0838.1.

IPCC 2014, *Climate Change 2014: Synthesis Report. Contribution of Working Groups I, II and III to the Fifth Assessment Report of the Intergovernmental Panel on Climate Change*, https://doi.org/10.1017/CBO9781107415324.

Kelman, I 2014, 'No change from climate change: vulnerability and small island developing states', *The Geographical Journal*, https://doi.org/10.1111/geoj.12019.

Osbahr, H, Twyman, C, Adger, WN, Thomas, DSG, 2010, 'Evaluating Successful Livelihood Adaptation to Climate Variability and Change in

Southern Africa', *Ecology and Society*, https://doi.org/10.5751/ES-03388-150227.

Pelling, M, Uitto, J 2001, 'Small island developing states: natural disaster vulnerability and global change: Global Environmental Change Part B', *Environmental Hazards*, https://doi.org/10.1016/S1464-2867(01)00018-3.

Pielke, R, Rubiera, J, Landsea, C, Fernández, M, Klein, R 2003, 'Hurricane Vulnerability in Latin America and The Caribbean: Normalized Damage and Loss Potentials', *Natural Hazards Review* 4, 101–114.

Planning Institute of Jamaica 2017, *Review of Economic Performance*, January–March 2017. PIOJ: Kingston, Jamaica.

Poncelet, JL 1997, 'Disaster management in the Caribbean', *Disasters*, https://doi.org/10.1111/1467-7717.00061.

Pulwarty, RS, Nurse, LA, Trotz, UO 2010, *'Caribbean Islands in a Changing Climate. Environment: Science and Policy for Sustainable Development.'* https://doi.org/10.1080/00139157.2010.522460.

Scandurra, G, Romano, AA, Ronghi, M, Carfora, A 2018, 'On the vulnerability of Small Island Developing States: A dynamic analysis', *Ecological Indicators.* https://doi.org/10.1016/j.ecolind.2017.09.016.

Smit, B, & Wandel, J 2006, 'Adaption, adaptive capacity and vulnerability', *Global Environmental Change*, https://doi.org/10.1016/j.gloenvcha.2006.03.008.

Spence, B, Katada, T, Clerveaux, V, 2005, *Experiences and Behaviour of Jamaican Residents in Relation in Hurricane Ivan*. Tokyo: Japan International Cooperation Agency.

Taylor, DA, Kuwornu, JKM, Anim-Somuah, H, Sasaki, N 2017, 'Application of livelihood vulnerability index in assessing smallholder maize farming households' vulnerability to climate change in Brong-Ahafo region of Ghana', *Kasetsart Journal of Social Sciences*, https://doi.org/10.1016/j.kjss.2017.06.009.

Taylor, MA, Jones, JJ, Stephenson, TS 2016, 'Climate change and the Caribbean: Trends and implications' in E. Thomas Hope (ed.) *Climate Change and Food Security: Africa and the Caribbean*, Earthscan, 31-56. https://doi.org/10.4324/9781315469737.

LIST OF CONTRIBUTORS

Dr. Clinton Beckford is Associate Professor in the Faculty of Education, University of Windsor, Ontario, Canada. He is a Geographer with a Ph. D in Geography from the University of the West Indies. He is a researcher and scholar in the area of agriculture, food, and food security with a particular focus on tropical small-scale farming systems. Dr. Beckford's research also focuses on local food production systems within the context of a food sovereignty framework. His current research explores urban food production systems including home and back yard gardening and community gardens in Canada as well as the social, economic and environmental roles of food rescue programmes. He is the author/editor of two books on agriculture and food security in the Caribbean, *Globalization, Agriculture and Food in the Caribbean. Climate Change, Gender and Geography*, 2016 and *Domestic Food Production and Food Security in the Caribbean: Building Capacity and Strengthening Local Food Production Systems* 2013.

Dr. Noureddine Benkeblia is a Professor of Crop Science and Head of the Laboratory of Crop Science, Department of Life Sciences, Faculty of Science and Technology, at the University of the West Indies, Mona Campus, Kingston, Jamaica. He is also Director and Coordinator of the Agricultural Programmes at UWI-Mona. His main research areas are (i) environmental stresses and physiology and biochemistry of crops, (ii) pre- and postharvest metabolomics of fresh crops, and (iii) postharvest metabolism of fresh crops. Dr. Benkeblia published over 200 publications including books, chapters and research papers.

Dr. Andreas Berk is an agricultural scientist. Hi doctorate degree is in poultry nutrition. He works at the Institute of Animal Nutrition, Friedrich-Loeffler-Institut (FLI), Federal Research Institute for Health (former FAL). His fields are pig nutrition and feed evaluation.

Dr. Donovan Campbell, is a lecturer in the Department of Geography and Geology at the University of the West Indies, Mona Campus in Kingston, Jamaica. Dr. Campbell works in the area of agricultural geography with special interest in climate change impacts on agriculture and food security, and the prospects for building capacity in climate change adaptation and resilience in small island development states. Dr. Campbell's work and growing reputation in the area of climate change has been recognized by the Intergovernmental Panel on Climate Change IPCC assessment, the leading authority on climate change globally.

Dr. Gerhard Flachowsky is a retired professor form Institute of Animal Nutrition, Friedrich-Loeffler-Institute, Braunschweig, Germany. He is an expert in agricultural science, with specializations in poultry and beef nutrition. He has worked at various teaching and research institutes all over the world, including Ethiopia, Norway, UK and the USA. He is a former head of the Institute of Animal Nutrition of the Federal Research Centre of Animal Health in Germany. He is currently a member of the Panel for Feed Additives (FEEDAP) of the European Food Safety Authority (EFSA).

Dr. Ingrid Halle expertise is in the area of agriculture and animal nutrition. She has worked at the Institute for Small Animal Research of the FAL and since 2001 at the Institute of Animal Nutrition (FLI). She works in the fields of poultry, pet and exotic animal nutrition, with a special focus on feed stuffs, feed additives (trace elements - Iodine, Selenium; vitamins, essential oils, rare earth, enzymes), metabolism studies, feed safety, GMO-plants (corn, potatoes), and acrylamide.

Dr. Blessing Igbokwe is a high school Geography teacher with the Windsor-Essex County District School Board, and Sessional Instructor at the Faculty of Education, University of Windsor. Her research includes environmental education and the role of GIS technologies in the exploration of geographical questions.

Dr. Ulrich Meyer is Deputy Head of the Institute of Animal Nutrition, Friedrich-Loeffler-Institut (FLI), Federal Research Institute for Health in Braunschweig, Germany where he leads the working group on cattle nutrition. He is an agricultural scientist and holds a doctorate degree in animal nutrition. His research interests focus on the nutrition of dairy cows and growing cattle. An important part of his research is the iodine supply of beef and dairy cattle and the carry over associated with the carry-over of iodine from feed into food of animal origin.

Dr. Hugh Semple, holds a PhD in Geography from the University of Manitoba, Canada. He is currently a professor of Geography in the Department of Geography and Geology, Eastern Michigan University where he teaches courses in GIS and geography. His research interests are (a) disease mapping and spatial analysis of public health data, (b) food production and food accessibility issues in the Caribbean and the USA, and (c) urban revitalization in the Caribbean and the USA.

INDEX

A

access, x, xii, xiii, xvi, xvii, 21, 59, 60, 76, 77, 80, 91, 160, 163, 165, 168, 170, 171, 177, 180, 181, 182, 184, 185, 188, 196, 208, 210, 211, 213, 216
accessibility, 72, 78, 170, 225
accretion, 99
acid, 5, 6, 8, 9, 18, 20, 134, 135, 136, 138, 139, 141, 144, 145, 146, 152, 153, 158
acidity, 5, 39, 135, 137, 138, 140, 143, 144
adaptation, xi, xviii, 190, 193, 196, 199, 200, 214, 215, 216, 224
adaptive capacity, xvii, xx, 190, 194, 195, 199, 200, 201, 202, 205, 215, 218, 221
additives, 49, 51, 84, 116, 117, 121, 126, 224
adults, 31, 33, 35, 39, 106, 120, 184
Africa, xii, xv, 2, 111, 116, 161, 221
African American, 70, 167, 185, 188
African Americans, 62, 167, 185, 188
age, 11, 16, 32, 98, 113, 174, 199, 206, 214, 215
agencies, 198, 214
aggregation, 206, 210

agricultural market, 210
agricultural producers, 216
agricultural techniques, 174, 177
agriculture, v, ix, x, xi, xii, xiv, xviii, 79, 162, 163, 166, 172, 182, 183, 186, 187, 190, 191, 193, 194, 214, 215, 216, 217, 219, 220, 223, 224
agro-ecological, xi, xvii, 189, 194
agroforestry, xv, 2, 22
agro-processed, 151
albumin, xxii, 95, 96
algae, 45, 47, 54, 120, 137
Amartya Sen, x, 171
amino acid, 45, 99
ammonia, 148
analysis, x, xxiii, 24, 34, 56, 57, 61, 63, 64, 77, 80, 102, 116, 128, 169, 192, 193, 194, 195, 198, 199, 201, 212, 218, 220, 221, 225
Animal(s), xv, xvi, xxii, 10, 29, 30, 31, 32, 34, 35, 36, 38, 45, 46, 49, 50, 51, 53, 55, 81, 82, 84, 88, 89, 92, 97, 98, 99, 103, 106, 110, 112, 116, 117, 121, 123, 127, 129, 224, 225
Ann Arbor, 61, 62, 63, 78

Annotto Bay, 198, 200, 201, 210, 211, 216
antagonists, xix, 32, 86, 97, 105, 112, 157
antioxidant, 6, 23, 141, 153, 155, 156
antipyretic, 139
apples, 180
aroma, 7, 15, 16, 19
ascorbic acid, 18, 138, 144, 145, 146, 153
Ascorbic acid, 6, 141
Asia, xv, 1, 2, 7, 10, 27, 34, 43, 95, 132, 133, 161, 162
assessment, xvii, 23, 38, 79, 190, 194, 195, 198, 199, 200, 214, 216, 217, 224
assets, 195, 196, 200, 211, 212, 216
atmosphere, 19, 26, 27, 140, 144, 145, 146, 151, 157
attitudes, 181, 184, 195, 199
Australia, xv, 2, 22, 47, 133, 150, 152, 162, 164, 172, 181
Austria, 55, 83, 106, 127
automata, 70, 71, 75
awareness, xvii, 164, 196, 198, 200, 210, 211, 217

B

Bangladesh, 17, 18, 22, 23, 27, 77, 153, 156
banks, 163, 165, 166, 169, 170
barriers, 169, 179, 184
base, xx, 3, 61, 62, 63, 66, 70, 110, 182, 195, 208, 215
Bath, 90, 114, 198, 200, 201
beef, 108, 110, 224, 225
benefits, xiii, xv, xvii, 2, 38, 163, 168, 170, 180, 182, 184
beverages, 31, 55, 125, 150
bioavailability, 9
biochemical, 7, 21, 30, 54, 56, 127
biochemistry, xv, 150, 224
biosynthesis, 147
bleeding, 10, 139
blood, 56, 121, 126, 127

body weight, 32, 34, 109, 129
Bosnia, 89, 116
Bosnia-Herzegovina, 89
botanical, xv, 2
bottom-up, 193
brain, xv, 46, 49, 117
breed, 91, 97, 103, 105
British Geological Survey, 36, 37, 47, 50, 118
Brno, 122, 124, 129
broilers, xxiii, 97, 107, 109, 110, 126
bromine, 103
brown layers, 98
browning, xix, 142, 145, 146, 147, 148, 149
by-products, xxi, 21, 42, 43, 103, 114

C

cabbage, 177, 179
calcium, xxii, 49, 84, 85, 86, 87, 116, 117
California, xi, 24, 77, 79, 80, 131, 156, 164, 172, 186, 187
CA-Markov, 70, 71, 72, 74
campaigns, 21
Canada, xiv, xviii, 79, 99, 159, 160, 161, 164, 167, 171, 173, 175, 180, 185, 188, 223, 225
cancer, 168, 187
canned fruit, 8
capacity, 23, 195, 199, 200, 201, 216, 217, 218, 224
Carambola, viii, xv, xix, 131, 132, 133, 135, 137, 142, 144, 150, 152, 153, 154, 155, 156, 157, 158
carbohydrate, 5, 9, 26
carbon, 19, 140, 155
carbon dioxide, 19, 140, 155
cardiovascular disease, 181, 184
Caribbean, xv, 1, 2, 8, 132, 133, 134, 150, 184, 190, 191, 192, 206, 216, 219, 220, 221, 223, 225

Caribbean countries, 191
Caribbean Islands, 221
carotene, 5, 8, 138, 141
carotenoids, 138, 141
carry-over, 31, 55, 92, 93, 123, 126, 225
case study, 22, 160, 167, 172
cattle, 10, 84, 91, 92, 113, 121, 225
CDC, 77, 159, 184
Census, xxi, 62, 72, 80
census tracts, 61, 63, 64, 65, 66, 67, 69, 71, 75
cereal, xxi, 41, 42
challenges, ix, x, 172, 182, 193, 212, 213, 216, 217, 218
characteristics, 14, 15, 23, 27, 56, 61, 127, 138, 151, 153, 155, 168, 186, 188, 195, 199, 201, 208, 215, 218
chemical, xi, 7, 10, 26, 51, 83, 94, 115, 138, 151, 155, 157, 171, 179, 181
chemical characteristics, 151, 155
children, 34, 36, 38, 46, 51, 52, 53, 58, 121, 166, 187
Chile, 40, 41, 43, 44, 83, 108, 110
chilling, xix, 144, 147, 148, 149, 151, 155, 157
China, 14, 36, 52, 53, 80, 133
chromatography, 7, 25, 26
chronic diseases, xiv, xvii, 160, 168, 181
cities, 63, 80, 162, 163, 164, 166, 170, 185
climate change, xi, xiv, xvii, xviii, 165, 190, 191, 192, 193, 194, 195, 196, 199, 210, 211, 214, 217, 218, 219, 220, 221, 224
climate resilience, xvii, 190
climate variability, xi, xiv, xvii, 190, 192, 193, 195, 216
climates, 2, 24, 132, 136, 155
climate-smart agriculture., 214
coatings, 19, 28, 144, 153
color, 8, 11, 15, 16, 20, 152
colour, 135, 136, 140, 141, 142, 143, 144, 145, 146, 147, 148
colour break, 140, 141, 142

combined effect, 212
combustion, 123
commercial, xiv, xv, 2, 19, 21, 62, 99, 104, 113, 136, 141, 144, 150, 157
Commonwealth of Independent States, 56
communication, 179
Community, v, xvi, xvii, xx, xxiii, 61, 76, 79, 159, 160, 161, 162, 163, 164, 165, 166, 167, 168, 169, 170, 171, 172, 173, 174, 178, 180, 181, 182, 183, 184, 185, 186, 188, 190, 193, 194, 195, 196, 198, 199, 200, 201, 203, 204, 205, 206, 208, 210, 211, 212, 213, 214, 215, 216, 217, 218, 223
    gardens, xvi, 159, 160, 161, 162, 163, 164, 165, 166, 167, 168, 169, 170, 171, 172, 173, 178, 180, 181, 182, 183, 184, 185, 186, 188, 223
    groups, 209, 210, 216, 217
    social programs, 165
    support, 208
community-based, xvii, 165, 182, 193, 194, 214, 215, 216
    food, 165, 182
complexity, xii, 194
composite, 3, 144, 151, 194, 199, 200, 206, 211, 212
composition, xv, xxi, xxii, xxiii, 2, 5, 6, 19, 25, 26, 27, 56, 91, 93, 94, 113, 116, 128, 129, 137, 138, 139, 141, 145, 147, 152, 155, 158, 188
compounds, xix, 6, 7, 10, 19, 25, 49, 52, 116, 117, 138, 141, 155, 158
concentrations, xxii, 31, 36, 38, 85, 91, 92, 95, 96, 103, 107, 113, 115, 120, 125, 126, 128, 143
conceptualization, 195
consumers, xvii, xxiii, 5, 39, 106, 168, 182, 184
consumption, xii, xiii, xvii, 5, 32, 34, 39, 44, 57, 82, 106, 110, 160, 163, 169, 171, 180, 181, 186

containers, 177, 178
contamination, 18, 91
convenience stores, 60, 76, 169
conventional milk, 89, 92, 114, 125
cooked, 8, 9, 15, 20
cooking, 8, 9, 10, 15, 186
cooperative activities, 215
coping strategies, 195
corner shops, 169
correlation, 36, 91, 96
correlation coefficient, 96
cost, xiii, 172, 196
Cow(s), xxii, 32, 35, 49, 51, 55, 82, 84, 86, 87, 89, 91, 113, 114, 115, 117, 118, 120, 121, 122, 123, 124, 127, 128, 225
crambe meal, xxii, 87
crops, 1, 26, 28, 131, 143, 156, 174, 175, 177, 224
cultivars, 4, 7, 15, 25, 134, 135, 136, 137, 138, 142, 145, 154
cultivated, 1, 2, 3, 14, 131, 132, 133, 167, 173, 174, 179, 180
cultivation, 23, 167, 171, 175, 213
cultivation techniques, 175
cultural values, 166, 177
culture, 160, 163, 174
Czech Republic, 88, 89, 90, 107, 122

D

dairy, xxii, 32, 35, 48, 84, 87, 91, 92, 113, 114, 115, 116, 118, 121, 123, 124, 125, 127, 128, 225
data, 34, 35, 36, 38, 42, 61, 62, 65, 66, 67, 72, 75, 77, 94, 96, 97, 107, 110, 132, 166, 167, 173, 179, 195, 198, 199, 200, 201, 225
data analysis, 195, 198, 199
data collection, 173, 195
decay, 11, 144
deciduous, 2, 4, 134

deficiency, x, xiv, xvi, xviii, 30, 31, 38, 39, 46, 49, 57, 113, 211
deficit, x, xii, xiii, xiv, xvii, 182
demand, xvii, 5, 18, 19, 34, 106, 182, 206, 218
demographic factors, 196, 217
Denmark, 31, 36, 37, 54, 55, 83, 90, 125
Department of Agriculture, 22, 28, 78, 188
dependency ratio, 206, 214, 215
depolymerization, 143, 144
deposition, 95, 105, 126
desiccation, 137, 146, 148
developed countries, x, xiv, 164, 165
developing countries, xiii, xiv, 165
diabetes, xiii, 10, 60, 77
dietary intake, 48, 163, 167
dietary iodine, 58, 96, 118, 121, 123, 124, 127
diets, xix, xxiii, 1, 84, 86, 93, 101, 103, 104, 116, 122, 126, 131, 187
digestion, 123
disaster, 191, 210, 211, 214, 216
diseases, xi, xiii, xvii, 10, 13, 14, 144, 149, 150, 154, 160, 168, 181, 212
disinfectants, 90
disinfection, 38, 47, 89
distribution, x, xii, 5, 22, 56, 64, 68, 119, 163, 171
diversification, 215, 219
diversity, xi, xiii, 174, 185, 206
Dolphin Head, 198, 200, 201, 208, 210, 217
dosage, 97, 98, 103
dose-response, 31, 36, 95, 97, 109, 127
drinking water, 31, 35, 36, 37, 38, 47, 50, 51, 52, 53, 54, 55, 125
droughts, 137, 191, 192, 212, 213, 216, 217, 220
dry matter, 32, 35, 40, 45, 84, 91
ducks, 98, 107, 113
duckweed, 45

E

economic, ix, x, xiii, xv, xvii, xx, 2, 13, 60, 61, 63, 74, 75, 76, 152, 182, 188, 190, 191, 193, 196, 206, 209, 213, 214, 223
economic development, 76, 132, 190
Economic Research Service, 78
ecosystem, 193, 206, 213, 214, 215
ecosystem services, 193, 206, 213, 214, 215
eczema, 139
edible, 4, 5, 6, 9, 19, 20, 28, 53, 56, 97, 138, 139, 144, 151, 153
education, 21, 63, 76, 164, 187, 206, 215, 225
Egg(s), xxii, xxiii, 30, 31, 55, 56, 81, 82, 93, 94, 95, 96, 97, 98, 99, 100, 101, 103, 104, 105, 106, 112, 113, 115, 117, 118, 119, 122, 124, 125, 126, 127, 128, 129, 130, 179
  white, xxii, 93, 95, 96, 104
  yolk, xxii, 31, 93, 95, 96, 98, 122
Egypt, 94, 120
Elam, 183
elementary school, 187
emergency, 171
employment, xv, 19, 74, 163, 191
employment levels, 74
employment opportunities, 19
entitlements, x, 171, 180
environment, xvii, 25, 166, 182, 187, 193, 218
environmental, xi, xiii, xvii, 35, 163, 178, 182, 190, 193, 212, 218, 219, 223, 224, 225
environmental change, 218, 219
environmental issues, 190
environmental stress, 35, 224
environments, 137, 164, 165
enzymes, 148, 151, 153, 225
Ethiopia, 39, 55, 224
ethnicity, 62, 77
ethylene, 140, 142, 146, 152, 154, 157
Europe, 1, 11, 34, 47, 49, 56, 57, 99, 113, 116, 131, 161, 164
European Food Safety Authority, 29, 82, 106, 224
European Union, 84, 107, 117
evidence, v, xi, 72, 140, 168, 170, 187, 217
evolution, 63, 78
excretion, 36, 85, 93, 118
exercise, 178, 180
experimentation, xxii, xxiii, 97, 98, 100, 101, 104
expertise, 166, 217, 224
exposure, xvii, xx, 48, 60, 115, 146, 147, 148, 160, 190, 194, 199, 200, 201, 202, 204, 210, 212, 213
extraction, 8, 19, 158
extracts, 20, 50

F

factor maps, 66, 70, 71, 72, 75
families, 63, 161, 167, 173, 174, 177, 180, 181, 213
family members, 174
famine, 171, 188
farm management, 84, 112
farmers, xiii, xv, 2, 18, 170, 172, 174, 182, 186, 187, 188, 190, 199, 213, 216, 217, 218, 220
farming, xiv, xvii, xxii, 89, 90, 92, 161, 178, 181, 189, 194, 199, 211, 215, 216, 217, 218, 219, 221, 223
farms, 84, 89, 91, 104, 107, 115, 126, 129, 179
fast food, xvi, 60, 75, 168, 182
fatty acids, 22
feed additives, 49, 116, 117, 224
feed-conversion, 103
feeds, xxi, 30, 35, 41, 86, 95, 105, 126
field studies, 39, 107

films, 144, 145
financial, 170, 172, 211, 212, 218
financial capital, 218
financial resources, 172, 212
Finland, 36, 83, 94, 108, 110
firmness, 140, 143, 144, 145, 146, 147
fish, xxiii, 39, 46, 81, 82, 110, 111, 112
Flagaman, 198, 200, 201, 208, 211, 212, 216
flavonoids, 6, 11, 141
flexibility, 196, 200
flocks, 91, 104, 106
flooding, 137, 153, 213, 216
Florida, 2, 24, 25, 131, 132, 133, 134, 136, 152, 153, 154, 155, 156, 157
flour, 5, 9, 20, 27
flower(s), xix, 3, 4, 9, 20, 134, 135, 137, 150, 177
Food, v, ix, x, xi, xii, xiii, xiv, xv, xvi, xviii, xxii, xxiii, 7, 9, 10, 18, 19, 21, 22, 27, 30, 31, 32, 34, 36, 45, 46, 47, 50, 51, 52, 54, 55, 57, 59, 60, 61, 62, 63, 64, 66, 67, 70, 71, 72, 74, 75, 76, 77, 78, 79, 80, 81, 82, 83, 93, 94, 110, 112, 113, 115, 116, 117, 119, 120, 121, 123, 125, 130, 132, 139, 150, 156, 160, 161, 162, 163, 164, 165, 166, 167, 168, 169, 170, 171, 172, 173, 174, 177, 178, 179, 180, 181, 182, 183, 185, 186, 187, 188, 191, 192, 213, 217, 219, 223, 224, 225
  banks, 165, 166, 169, 170
  choices, xiii, 170, 181
  deserts, x, xiii, xvi, xviii, 59, 60, 61, 62, 63, 64, 66, 67, 70, 71, 72, 74, 75, 76, 77, 78, 80, 163, 167, 168, 170, 172, 181, 182, 186, 188
  insecurity, xiv, xv, xvi, 21, 59, 163, 171, 180, 217
  justice, 166
  preservation, 173
  security, ix, xi, xii, xiii, xiv, xvii, 19, 150, 162, 165, 166, 169, 170, 171, 172, 173, 180, 182, 185, 192, 217, 223, 224
  stamp, 165, 169
  systems, 166, 170, 182, 183
food additive, 51, 121
Food and Agricultural Organization, 162, 185, 220
food chain, 47
food industry, 30
food intake, 180
food production, x, xi, xii, xiii, xiv, xviii, 161, 163, 164, 165, 166, 172, 181, 182, 223, 225
food products, 31
food security, ix, xi, xii, xiii, xiv, xvii, 19, 150, 162, 165, 166, 169, 170, 171, 172, 173, 180, 182, 185, 192, 217, 223, 224
forages, xxi, 40, 42
formal education, 215
formal training, 206, 215, 217
formula, 68, 69
framework, xiv, 171, 194, 195, 201, 223
France, xiii, 83, 90, 114, 133
freezing, 7, 20, 24, 27
fresh fruits, xvii, 17, 76, 159, 160, 168, 170, 173, 181
freshwater, 110, 111
fructose, 5, 6, 141
Fruit, xiv, xv, xvi, xix, xxi, xxiii, 1, 4, 5, 6, 8, 9, 11, 12, 13, 14, 15, 16, 17, 18, 19, 20, 21, 22, 23, 24, 26, 27, 28, 32, 132, 133, 134, 135, 136, 137, 138, 139, 140, 141, 142, 143, 144, 145, 146, 147, 148, 149, 150, 151, 152, 153, 154, 155, 156, 157, 158, 163, 169, 174, 180, 181, 184, 186, 187
  salads, 150
functional food, 156

G

gardens, xvi, 76, 80, 160, 161, 162, 163, 164, 165, 166, 167, 168, 169, 170, 171, 172, 178, 181, 182, 184, 186, 187
gender, 32, 167, 199, 215
gene expression, 152
genetic diversity, 26
genotypes, 98
Geographic Information System (GIS), 70, 80, 201, 225
geography, 66, 224, 225
Germany, xiii, xiv, xxi, 29, 34, 35, 37, 39, 40, 41, 42, 43, 51, 55, 81, 83, 89, 90, 107, 108, 110, 111, 112, 114, 118, 119, 125, 126, 127, 185, 224, 225
global, v, ix, x, xi, xii, xiii, xiv, xviii, 36, 45, 112, 113, 162, 163, 192, 221
global climate change, xi
globalization, xii, 60, 219
glucose, 6, 141
glucosinolates, xxii, 35, 86, 87, 88, 99, 103, 112, 114
goats, 91, 92, 122, 124, 129
goiter, xvi, 36
goitrogens, 84
government, 161, 162, 209, 210, 214
grass, xxi, 40, 214
grassroots, 161, 193
Greater Antilles, 191
green fruit, 15, 143
Green Revolution, ix, xi
grocery store, 168, 182
groundwater, 36, 37
growth, ix, xvi, 13, 15, 16, 24, 26, 27, 63, 70, 74, 123, 137, 162
Guyana, 133, 191

H

harvest, 12, 14, 16, 17, 18, 21, 138, 140, 141, 149, 151, 156, 157, 173, 180
harvesting, 13, 15, 16, 132, 137, 142, 143, 149
Hawaii, 132, 133, 136, 158
hazards, 192, 199, 212, 216
health, ix, xiii, xvi, xvii, 19, 27, 30, 32, 39, 45, 46, 48, 60, 61, 76, 99, 114, 132, 160, 162, 163, 164, 168, 171, 180, 181, 182, 183, 184, 186, 187, 188, 213, 214, 225
health risks, 168, 181
heavy metals, 45, 54
hens, 32, 55, 97, 98, 102, 103, 104, 106, 107, 115, 127, 128
heritability, 103
heterogeneity, 199
history, 43, 64, 80, 132, 160
hormone, xv, 29, 38, 49, 85, 99, 103, 116, 117, 126
horticultural, 142, 156
horticultural crops, 156
household income, 171, 180
Household(s), xvii, 10, 30, 39, 52, 160, 162, 163, 165, 166, 167, 168, 169, 170, 171, 172, 173, 180, 181, 182, 190, 195, 196, 199, 206, 210, 213, 218, 221
human brain, 30
human capital, 218
human dimensions, 193
human health, ix, xiii, xvii, 27, 32, 45
human milk, 82, 122
Human(s), ix, x, xii, xiii, xv, xvi, xvii, xxi, 5, 27, 29, 30, 31, 32, 33, 35, 38, 44, 45, 46, 50, 51, 53, 81, 82, 93, 103, 106, 107, 110, 112, 113, 122, 125, 193, 218
 milk, 82, 122
 systems, 193
humidity, 18, 39, 143, 144
hunger, ix, x, xiii, xiv, 113, 164, 171, 188

hurricane, 191, 192
hyperthyroidism, xvi, 30
hypertrophy, 99
hypothesis, 66, 69
hypothyroidism, xvi, 30

I

Iceland, 83, 89
I-content, xxiii, 31, 38, 90, 93, 94, 96, 98, 99, 103, 109, 112
immigrants, 167, 174
improvements, 2, 54, 124, 132
incidence, 144, 148
income, xv, xvi, xvii, 2, 18, 19, 21, 25, 27, 47, 59, 60, 63, 66, 70, 76, 77, 80, 160, 161, 163, 164, 167, 168, 169, 170, 171, 180, 181, 182, 184, 186, 193, 215
incompatibility, 154
increased access, 160, 180
independence, 66, 67, 69
index, 16, 19, 194, 199, 200, 201, 218, 221
India, xv, 5, 8, 10, 11, 13, 17, 23, 37, 47, 55, 133, 139, 150, 154, 188
individuals, 160, 161, 164, 171, 180, 206, 210, 211, 214, 218
Indonesia, 11, 133
industrial, xiii, xiv, 54, 60, 62, 63, 79, 182
industrialized societies, xiii
industries, 19, 213
information, v, xviii, 9, 21, 76, 91, 190, 194, 199, 210, 211, 216
ingredient, 19
injury, xix, 144, 147, 148, 149, 150, 151, 155, 157
inner city, 60, 166, 186
inorganic, 99
insect pests, 13, 177
insecurity, xiv, xv, xvi, 21, 59, 163, 171, 180, 217
intakes, 34, 48, 58, 168

integration, xv, 195, 217
interaction, 63, 164, 193, 195
interference, 85, 99
intervention, 181, 194, 198, 217
intervention strategies, 217
interventions, 195, 214, 215, 217
interviews, 199
iodate, xix, xxii, 39, 49, 84, 85, 86, 87, 93, 98, 116, 117
Iodine, vii, xv, xxi, xxii, xxiii, 29, 30, 32, 34, 35, 40, 41, 43, 44, 45, 47, 48, 49, 50, 51, 52, 54, 55, 57, 81, 82, 84, 87, 89, 93, 94, 95, 96, 97, 98, 100, 101, 102, 104, 105, 106, 108, 109, 110, 113, 114, 115, 116, 117, 118, 119, 120, 121, 122, 123, 124, 125, 127, 128, 129, 130, 224
iodized salt, 31, 32, 39, 46, 48, 52
IPCC, 193, 201, 220, 224
I-recovery, xxii, 95, 96
I-requirements, 34, 38, 46
iron, 103, 115
irrigation, 137, 174
island, xvii, 22, 36, 133, 190, 191, 192, 193, 216, 220, 221, 224
I-sources, 32, 39, 46, 99
Israel, xv, 134, 150, 188
issues, v, ix, x, xi, xiii, xiv, xv, xvi, xvii, xviii, 2, 76, 132, 169, 177, 193, 215, 225
I-supply, 31, 93, 104
I-transfer, 99, 103

J

Ja REEACH, 195, 198
jackfruit, xiv, xv, xix, xxi, 1, 2, 4, 5, 6, 7, 8, 9, 10, 11, 12, 13, 14, 16, 17, 18, 19, 20, 21, 22, 23, 24, 25, 26, 27, 28
Jamaica, viii, xiv, xvii, 1, 8, 131, 189, 191, 192, 193, 194, 195, 198, 212, 215, 216, 217, 218, 220, 221, 223, 224
Japan, xiii, 34, 57, 221

Japanese, 43, 57
Junction, 198, 200, 201, 212, 217

K

kale, 85, 99
kidney, 109
kill, 159
knowledge, xiv, 45, 76, 163, 166, 170, 175, 178, 186, 187, 195, 200, 210, 211, 220

L

Lactating, 46, 92, 97, 113
　mammals, 97
landscape, 24, 76, 162, 183
Latin America, 132, 192, 221
laying hens, xxii, 30, 32, 55, 56, 93, 95, 96, 97, 99, 106, 107, 113, 118, 122, 124, 126, 127, 130
laying performance, 98, 126
leadership, 166, 216
learning, 164, 195, 196, 200
leaves, xix, 3, 9, 10, 15, 134, 135, 139, 151
legumes, xxi, 42, 43, 164, 171
lesions, 142, 149
leukocytes, 121
level, xi, xii, xxii, xxiii, 2, 18, 31, 32, 34, 35, 46, 52, 55, 56, 74, 92, 97, 100, 101, 103, 104, 126, 136, 162, 172, 173, 192, 193, 198, 199, 200, 201, 206, 208, 210, 211, 212, 213, 214, 215, 216, 217, 218
level of education, 206
light, xi, 15, 39, 85, 135, 140, 142, 149, 179
liquid chromatography, 156, 158
Livelihood, xvii, 190, 191, 194, 195, 196, 199, 200, 201, 206, 208, 210, 211, 212, 213, 215, 217, 218, 221
　diversity, 206
　strategies, 199, 200
liver, 108, 109, 110

livestock, 10, 56, 91, 112, 163, 211
local community, 173, 182
local government, 161
local knowledge, 193, 195, 217, 219
location, 35, 66, 70, 72, 74, 163, 180, 181, 186, 198
low density polyethylene, 144
low temperatures, 148, 152
low-income, xvi, xvii, 47, 59, 60, 63, 160, 161, 163, 164, 167, 168, 171, 180, 181, 182, 184, 186

M

majority, xiii, 61, 174
Malaysia, xv, 5, 132, 133, 150, 158
malnutrition, xiii, 21
management, 84, 97, 112, 137, 184, 214, 219, 221
manufacturing, 60, 62, 63
maps, 67, 69, 70, 71, 72, 74, 75, 201
market conditions, 210, 211, 218
market prices, 17, 210, 211
marketability, 145, 149
marketing, 5, 17, 136, 141, 163, 213, 219
markets, xii, 17, 63, 142, 165, 170, 172, 182, 186, 188
Markov Chain, vii, xvi, 59, 61, 64, 66, 69, 70, 76, 78, 79
Maryland, 26, 173, 184
mass, xiii, 7, 25, 26, 103, 123, 156, 171
mass spectrometry, 7, 25, 26, 123, 156
materials, 19, 141, 142
matrix, 65, 66, 67, 69, 70, 71, 72, 75, 199, 200
matter, xxii, 11, 94, 96, 107, 110, 133
maturation, 15, 16, 140
maturity, 5, 11, 12, 13, 14, 15, 16, 17, 18, 26, 134, 137, 140, 141, 144, 145, 151, 155, 157, 179

maximum, 31, 34, 35, 38, 47, 106, 107, 142, 192
MCP, 19, 28, 147, 154
measurements, 12, 13, 32, 107
meat, xxiii, 8, 20, 32, 46, 59, 60, 61, 81, 82, 107, 108, 109, 110, 120, 125, 128
median, 63, 66, 70, 167, 170
medicinal, 9, 153, 156, 177, 180
medicine, 21, 139
membership, 166
mental development, 30, 46
mental retardation, xvi
meta-analysis, 128
metabolic, 34
metabolism, 103, 224, 225
metaphor, 60, 65
methodology, 194, 195, 198
Michigan, vii, xvi, 59, 61, 62, 66, 78, 79, 188, 225
micro-climate, 216
micronutrients, xii, xiv, 82
micro-scale, xvii, 190, 193, 215
milk, xix, xxii, 8, 30, 31, 39, 48, 49, 51, 55, 59, 60, 81, 82, 83, 84, 85, 86, 87, 88, 89, 90, 91, 92, 93, 107, 110, 113, 114, 115, 116, 117, 118, 120, 121, 122, 123, 124, 125, 126, 127, 128, 129, 169
Mitchell Town, 198, 200, 201, 212, 214, 215
mitigation, 196, 199, 210
mixed methods research, 194
Model(s), 24, 61, 64, 66, 67, 70, 73, 75, 76, 78, 79, 129, 184, 191, 193
models, 184, 191, 193
modifications, 21, 151, 153
moisture, 18, 39, 177
muscles, 109, 110
mushrooms, 42
mycelium, 149

N

NAS, 45, 53
National Academy of Sciences, 45, 53
National Health and Nutrition Examination Survey, 184
national policy, 216
National Research Council, 30, 34, 53
natural disaster, 221
natural hazards, 191, 199
natural resources, 215
New Zealand, 47, 151
Nobel Prize, ix, x, 171
Nordic Nutrition Recommendations, 33, 53
North America, 161, 164
Norway, 83, 88, 89, 90, 120, 224
NRC, 30, 34, 53, 54
null, 66, 69
null hypothesis, 66, 69
nutrient concentrations, 103
nutrients, xv, 2, 57, 82, 99, 125, 130
Nutrition, xi, xiv, xv, xvi, xviii, 19, 21, 25, 29, 30, 33, 43, 45, 47, 48, 50, 51, 53, 56, 58, 76, 82, 93, 110, 112, 119, 120, 124, 125, 129, 132, 160, 163, 164, 165, 166, 171, 172, 177, 179, 180, 181, 185, 186, 187, 224, 225
  security, xiv, xvi, xviii, 21, 132, 160, 164, 165, 166, 172
nutritional status, xv, 2, 132
nuts, 27, 42

O

obesity, xiii, xvi, 60, 77, 160, 169, 170, 181, 186
oil, 8, 9, 10, 19, 20, 144
open spaces, 183
operations, 17, 62
opportunities, 159, 181, 217

organic, 6, 89, 91, 99, 110, 114, 124, 125, 128, 171, 181
organic food, 182
organization, xx, 165, 170, 196, 207, 208, 210
oxalic acid, 6, 134, 135, 138, 158
oxidation, 85
oxygen, 19, 85, 145, 146, 157

P

Pacific, 2, 191
packaging, 17, 19, 26, 27, 140, 145, 151, 152, 154
participant observation, 199
participants, 166, 167, 181, 199
participatory, 195, 199, 216
pathogens, 13, 89
perceptions, 195, 199
perianth, 5, 20
perishability, 21
perishable, 4, 143
peri-urban, 160, 161, 162, 163, 164, 183
pests, xi, 13, 14, 177, 212
pH, 3, 136, 140, 144
phenolic compounds, 138, 158
Philippines, 10, 18, 22, 24, 25, 133
physical activity, 53, 159
physical characteristics, 23, 138
physicochemical properties, 22, 27
physiological, 21, 32, 148, 152, 153, 155, 157
physiological factors, 155
physiology, xv, 38, 125, 150, 155, 224
pigs, xxiii, 97, 107, 108, 109, 110, 114, 120, 127
pith, 10
plant, xv, xvi, xxi, 2, 6, 32, 41, 62, 81, 134, 136, 153, 212
plant growth, 136

plants, 3, 16, 40, 45, 46, 85, 87, 153, 156, 157, 177, 178, 179, 180, 225
platform, 217
PM, 25, 27, 114
Poland, 83, 88
policy, v, 25, 61, 75, 76, 162, 165, 215, 216
policy initiative, 75
policy makers, 61, 75, 76, 218
politics, xi, 182
pollination, 137, 154
polyamine, 152
polymers, 140
polysaccharides, 140
polystyrene, 145
population, ix, x, xi, 2, 48, 62, 70, 92, 115, 128, 165, 168, 170, 173, 180, 191, 206, 214, 215
population density, 170
population growth, x
pork, 107, 118, 121
postharvest, xv, 2, 21, 24, 132, 143, 144, 145, 146, 147, 148, 152, 156, 157, 224
post-harvest losses, 17, 18
potassium, xxii, 5, 84, 85, 86, 87, 92, 117, 124, 128, 130, 137, 145
potato(es), 8, 43, 85, 225
poultry, xxii, 32, 46, 55, 94, 107, 108, 112, 125, 126, 163, 224
poverty alleviation, xvii, 160, 182
poverty line, 63
pragmatic, 193, 194, 219
precipitation, 190, 217
pregnancy, xvi, 30, 33, 46
preparation, 8, 49, 76, 117, 178
preschool children, 121
preservation, 173, 178
principles, 26, 32, 194, 195
probability, 64, 65, 66, 67, 68, 69, 71, 72, 75
probability distribution, 64, 69
processed milk, 92

Index

processing, xv, 5, 16, 17, 19, 20, 23, 26, 92, 124, 145, 150, 151, 173
producers, xii, 18
productivity, 136, 142, 212, 217
protection, 13, 196
pruning, 137, 155, 160
public health, xvi, 30, 39, 48, 184, 225
public lands, 161, 162, 165
public parks, 159, 162
Puerto Rico, 191
pulp, xxi, 4, 5, 6, 8, 9, 12, 17, 23, 85, 135
PVC, 142, 145

Q

qualitative, 194, 195, 199, 200, 219
quality, xiv, 8, 16, 17, 19, 20, 21, 23, 24, 25, 28, 50, 61, 115, 122, 134, 136, 144, 146, 147, 149, 151, 152, 153, 154, 160, 163, 168, 172, 181, 198
quality assurance, 25
quality of life, 61
quantitative, 194, 195, 199
quantitative technique, 194, 195
Queensland, 152, 158
questionnaire, 199

R

race, 77, 167, 174
racial minorities, 167, 169
rainfall, 2, 136, 137, 212, 216
rape, xix, xxiii, 85, 86, 93, 101
rape seed, xix, xxiii, 86, 93, 101
rapeseed, 86, 87, 93, 99, 103, 118, 122
recommendations, v, 32, 33, 46
recovery, xxii, 95, 96
redevelopment, 184
reducing sugars, 18, 140
refractive index, 19

region, 36, 37, 38, 68, 74, 75, 76, 115, 133, 190, 191, 221
regions of the world, 32, 38
regression, 86
regression equation, 86
regulations, 75
relationship, x, 24, 76, 97, 99, 106
requirements, 30, 31, 32, 33, 34, 35, 38, 46, 49, 75, 81, 106, 107, 113, 117
research, v, xiv, xv, xvii, xviii, 2, 22, 29, 31, 45, 60, 63, 82, 84, 98, 113, 150, 160, 166, 167, 170, 173, 180, 182, 186, 193, 194, 195, 200, 218, 219, 223, 224, 225
resilience, xi, xvii, 165, 190, 193, 214, 216, 217, 224
resource management, 199
resources, xi, 27, 169, 170, 172, 211, 213, 216, 217
respiration, 140, 144, 147
response, xii, 31, 36, 84, 95, 96, 97, 99, 109, 114, 127, 152, 162, 163, 165, 214, 215
restaurants, 60, 75, 169, 185
retail, 17, 124, 128
Rhizopus, 14
Rio Minho, 198, 200, 201, 213, 217
Ripe, xix, 4, 5, 6, 7, 8, 10, 12, 15, 17, 18, 20, 134, 135, 138, 139, 140, 141, 142, 149
  fruit, xix, 4, 6, 7, 10, 15, 17, 20, 135, 139, 140, 141, 142, 149
risk, 31, 38, 82, 184, 187, 196, 210, 211, 213, 215, 216
room temperature, 8
root, 79, 137, 153, 214
root growth, 137, 153
routes, 60, 76
rubber, 10, 24
ruminants, 91, 97, 113
rural, xv, xvii, 2, 21, 25, 38, 59, 132, 160, 161, 162, 168, 186, 193, 196, 218, 219
  areas, 21, 59, 160, 161, 162
  development, 219

people, xvii, 25, 132
  population, xv, 2

S

safety, 49, 50, 116, 117, 166, 212, 225
safety-net, 212
saline, 36
salinity, xi, 137
salt-water, 110
School, 21, 38, 133, 162, 164, 170, 185, 187, 225
  gardening, 170
school community, 185
science, xiii, 49, 77, 114, 116, 118, 119, 123, 124, 126, 127, 128, 224
scientific data, 217
sea level, xi, 2, 136, 192
seaweeds, 32, 43, 45, 46, 50, 51, 53, 56
security, ix, xi, xiv, xvi, xviii, 21, 132, 160, 163, 164, 165, 166, 171, 172, 182, 190, 191, 223
seed, xxi, 4, 12, 13, 20, 27, 43, 105
selenium, 103, 123
senescence, 144, 147
sensitivity, xx, 194, 199, 200, 201, 202, 204, 213
serum, 118, 127
services, 63, 193, 196, 206, 208, 213, 214, 215
shade, 137, 142, 180
shape, xviii, 3, 132, 190
share, 165, 166, 173, 178
sheep, 91, 92, 126, 129
shelf-life, 18, 19, 26, 142, 143, 144, 147, 153, 154, 157
skin, 10, 15, 90, 134, 135, 145, 148, 149
skin diseases, 10
SLA, 195
Slovakia, 83, 88, 124
small business, 23

small-scale, xiii, 160, 161, 173, 174, 213, 219, 223
social capital, 218, 219
social fabric, 218
social interaction, 164
social network, 200, 208, 210
social programs, 165
social structure, 206, 218
social welfare, 169, 170
society, 125
Society of Nutrition Physiology, 34
socioeconomic, xvi, 61, 76, 132, 167, 173, 187, 190, 200, 213, 215, 218
sodium, 84, 144
software, 67, 70, 71
soil, 3, 42, 50, 53, 136, 153, 181, 190
solution, 90, 113, 170, 215
soup kitchens, 165, 166, 170
South Africa, 52
South America, 132, 133, 192
South Asia, 17
South Pacific, 133
Southeast Asia, 2, 132, 133
  Food, 165, 166, 223
  sovereignty, 165, 166, 223
Soybeans, 42
Spain, 83, 88, 89, 90, 125
spatial, 61, 69, 72, 74, 225
species, xviii, xxi, 1, 2, 32, 34, 35, 41, 43, 45, 49, 51, 92, 97, 107, 110, 111, 112, 113, 116, 117, 118, 131, 133, 177
Spring, 185, 198, 200, 206, 210, 211, 213, 214, 215
Spring Garden, 198, 200, 206, 210, 211, 213, 214, 215
Sri Lanka, 11, 23, 37, 42, 50
stability, 39, 52, 117, 130, 151, 196
stakeholders, 195, 198
standardization, 199
starch, 5, 9, 10, 22, 27
Starfruit, 132

state, xvii, xix, 21, 43, 64, 65, 66, 67, 68, 69, 70, 98, 220, 221, 224
statistical, 66, 67, 68, 199
Statistical Package for the Social Sciences, 199
stochastic, 64, 70, 76, 77, 80
stochastic model, 80
stochastic processes, 64, 77
stock, 10, 174
storage, 15, 17, 18, 19, 24, 26, 27, 28, 39, 85, 130, 136, 140, 142, 143, 144, 145, 146, 148, 149, 150, 151, 152, 153, 154, 155, 158
stress, 134, 212
stressors, 35, 210, 214
stroke, 168
structural barriers, 170
structure, 70, 79, 199
style, 20
Subsidies, 80
subsistence, 174
suburbanization, 60, 61
sucrose, 5, 6, 18, 141, 143
Sugar, 5, 8, 11, 20, 135, 141, 143, 144, 169, 213
cane, 213
summer feeding, 88
Supermarket(s), xvi, 60, 61, 63, 72, 76, 77, 80, 90, 92, 168, 169, 170, 171, 179, 180, 182
supplementation, xix, xxii, xxiii, 30, 39, 45, 84, 86, 88, 91, 92, 93, 94, 95, 96, 97, 98, 99, 100, 101, 102, 103, 104, 105, 106, 109, 112, 118, 121, 123, 124, 125
support services, 196, 208
susceptibility, xii, 137, 158, 168
sustainability, 182, 214
sustainable, xi, 171, 177, 181, 182, 195, 215, 217
sustainable development, 217, 221
sweet, 4, 7, 15, 20, 85, 132, 134, 135, 136, 152

sweetness, 16, 142, 145
Switzerland, 39, 41, 42, 44, 83, 94, 108, 111
symptoms, xiii, 137, 148, 149, 151
synthesis, 85, 91, 99, 219
systems, xv, xvii, 2, 172, 174, 182, 189, 194, 223

T

Taiwan, 133, 136
Tanzania, 164
target, 214
target population, 214
taste, 16, 19, 20, 21, 132, 145, 150, 171, 187
TCC, 141
teat-dipping, 89, 115
techniques, 143, 160, 174, 175, 177, 178, 181, 199
technological advances, x
technology, xiii, 18, 22, 147, 156, 196
Temperature(s), xi, 8, 14, 18, 19, 39, 92, 134, 142, 143, 145, 146, 148, 151, 152, 154, 155, 156, 157, 192
temporal variation, 188
terraces, 161, 163
TerrSet, 67, 70, 71, 73, 74, 80
texture, 8, 12, 19, 135, 145, 151
Thailand, 133, 136
thyroid, xv, 29, 34, 38, 46, 58, 85, 99, 116, 122, 126, 127
thyroid gland, xv, 30
thyroid stimulating hormone, 85
Timber, 11
toxicity, 47, 83, 94, 115, 121
trace elements, 55, 103, 127, 224
trade, xiii, 25, 26
traditional knowledge, 215
training, 196, 206, 215, 217
traits, 124, 130
transfer, x, 31, 55, 84, 85, 93, 105, 118, 125
transformation, 199

transition, ix, 63, 65, 66, 67, 68, 69, 70, 71, 72, 75, 161
transport, 16, 17, 166
transportation, xvi, 14, 17, 60, 79, 169
treatment, 19, 99, 139, 142, 145, 146, 147
Trees, 136, 137, 142
triiodothyronine, xv, 29
Trinidad, 8, 133, 191
Trinidad and Tobago, 8, 191
tropical fruits, xiv, xv, 1, 5, 131, 143, 148, 149, 151, 154
tropics, xv, xvii, 132, 150, 174, 182

U

unemployment rate, 62
UNICEF, 30, 31, 33, 37, 39, 57
uniform, 16, 146, 191
United Kingdom, xiii, 90
United Nations, 30, 162, 220
United States of America (USA), xiii, xiv, xv, xvi, 34, 37, 39, 44, 48, 62, 77, 80, 83, 94, 108, 111, 131, 132, 136, 150, 161, 164, 166, 167, 169, 173, 188, 224, 225
UPA, 162, 163
Urban, xvi, xvii, 59, 60, 61, 76, 79, 160, 161, 162, 163, 164, 165, 166, 167, 168, 169, 170, 172, 180, 181, 182, 183, 184, 185, 187, 188, 223, 225
  blight, 61, 160
  gardeners, 167
areas, xvi, 162, 163, 167, 168, 180, 182
population, 165, 168, 169
renewal, 160
urinary iodine, 36, 128
urine, 30, 34, 45, 51, 56, 112, 121, 125

V

vacuum, 145, 146, 154
validation, 71, 73, 75

value-added, 7, 25, 150
variables, 64, 71, 199, 206, 210, 211
variations, 36, 38, 115, 116, 141, 190
varieties, 5, 7, 8, 24, 136
Vegetable(s), xvi, xvii, xxi, 6, 8, 9, 12, 15, 16, 20, 24, 32, 42, 44, 76, 147, 159, 160, 161, 163, 164, 168, 169, 170, 171, 173, 174, 179, 180, 181, 183, 184, 186, 187, 188
vegetation, xxi, 40, 41
Vernamfield, 198, 200, 201, 208, 210, 211, 217
Vietnam, 133
violence, 160
vision, 164
visualization, 70, 71, 74
vitamin A, 5, 103
vitamin C, 6, 137, 145
vitamin D, 103
vitamins, 224
vulnerability, xiv, xvii, xviii, xx, 163, 180, 190, 191, 192, 193, 194, 195, 198, 199, 200, 201, 202, 203, 206, 210, 211, 212, 214, 215, 216, 217, 218, 220, 221

W

warming, xi, 190, 192
waste, 4, 9, 23, 187
water, xi, 8, 9, 31, 32, 35, 36, 37, 38, 42, 45, 46, 47, 50, 51, 52, 53, 54, 55, 56, 81, 110, 112, 119, 125, 134, 140, 143, 144, 145, 146, 148, 153, 156, 174, 213
water permeability, 146
weight, xiii, xxii, 4, 5, 9, 12, 13, 32, 34, 43, 75, 83, 93, 98, 104, 109, 129, 138, 140, 143, 144, 147
weight loss, 143, 144, 147
West Africa, 133
West Indies, 1, 131, 189, 223, 224
white layers, 98

winter keeping, 88
wood, 10, 11, 152
workers, 11
working class, 206
working population, 206
World Health Organization (WHO), xvi, 30, 31, 33, 34, 37, 39, 48, 51, 57, 82, 121
worldwide, x, 21, 39, 45, 57, 112, 113

Y

yeast, 124, 145

yield, 9, 15, 19, 35, 91, 114, 123, 142
yolk, xxii, xxiii, 31, 93, 94, 95, 96, 98, 104, 119, 122
youth, 165, 166, 170, 172, 181, 186
Ypsilanti, vii, xvi, 59, 61, 62, 63, 64, 66, 74, 75, 76, 78, 79, 80

Z

zinc, 103
zones, xvii, xx, 189, 194, 197